D0663769

When Y2K Dies

Virtually all Scripture references are quoted from the King James translation of the Holy Bible.

When Y2K Dies
Copyright ©1999 by Arno Froese
West Columbia, South Carolina 29170
Published by The Olive Press, a division of Midnight Call Ministries
Columbia, SC 29228 U.S.A.

Copy typist:	Lynn Jeffcoat
Proofreaders:	Angie Peters, James Rizzuti
Layout/Design:	James Rizzuti
Lithography:	Simon Froese
Cover Design:	J Spurling

Library of Congress Cataloging-in-Publication Data

Froese, Arno—Froese, Joel
    When Y2K Dies
    ISBN# 0-937422-45-2

    1. Computers—Y2K

All rights reserved. No portion of this book may be reproduced in any form without the written permission of the publisher.

Printed in the United States of America

This book is dedicated to the
Church of Jesus Christ worldwide.

It is intended to contribute toward a
better understanding of God's counsel
to men based on the Scripture.

The author/authors do not benefit
through royalties from the proceeds of
the sale of this book. All received funds
are reinvested for the furtherance of the
Gospel.

FOREWORD BY JERRY P. BROWN

When Arno Froese first told me he was writing a book on the "Y2K" situation, I cringed. "We don't need," I thought, "yet another book predicting doom and gloom and the destruction of society."

But then, I remembered who I was talking to. If you are expecting another "end of the world" supermarket tabloid book, close this book and put it back on the shelf. You are not going to get it.

Arno Froese has keen insight into the workings of society and the role God assumes in its guidance. Never one to engage in sensational writing, instead he relies on observation of behavior and how it has been foretold by Biblical prophecy.

In the coming months, you are going to be overwhelmed by articles and pundits trying to give their own interpretation on the coming millennium's computer "crisis."

I will now make my one and only prophecy concerning Y2K. Network TV's news programs will mention something about Y2K at least once in every broadcast. Go ahead, count them! Whenever there is a lack of hard news, you will hear about Y2K. I honestly believe that there is a concerted plan by the media to keep hyping this issue for no other reason than to keep an audience. The popular press is filled with warnings of impending gloom. What have you heard? Stock market crashes? Power blackouts? Worldwide economic crisis? Anyone can predict just about anything they wish. Remember, death and destruction sell newspapers. But, underneath this press manipulation, there are valid issues and concerns to discuss.

One of the most difficult problems concerns "embedded" systems, particularly in the power generation and medical equipment industries. But the best minds in the industry are working these problems out. Billions of dollars are being spent to resolve them. They are making tremendous strides. But then, it only takes one very serious television anchorman to take an isolated occurrence of problems to convince a million people that things are not well. Oh yes, you need to tune in again tomorrow to receive guidance on how to protect yourself. This is going to happen time and again. Go ahead, count them!

In embedded systems, the problem is overstated. The concern is actually in intervals, not specific dates. The book you are now reading is the first book I have read, written for the layman, that actually explains this in rational terms. You will read here that if there were problems with self-regulating computer controls, they would be occurring before or after January 1st. Hey, guess what? They ARE occurring. Up till now these problems have been labeled as computer glitches. But you sell more airtime today if you now call them Y2K "bugs." You will be inundated with these reports. Go ahead, count them!

Please take the time to read this book for a well balanced, Bible-centered view of the Y2K "crisis." You will be glad you did. You won't find sensational hype here.

Before I close, I will make one more prophetic statement. I know, I said I would make only one, but this one is important. Ready? Here it is: This year will pass. The days will follow each other patiently, as they have for the past millennium and the one before that. Right back to the beginning of Genesis, chapter 1. The impatience will come from the hearts of men, not from the heart of God. Do not fear this year and the date of January 1, 2000. It is just that, a date, a point in time.

—Jerry P. Brown,
Manager of Information Services
South Carolina Department of Labor, Licensing & Regulation

CONTENTS

INTRODUCTION BY ARNO FROESE

In late 1996, numerous books and articles began appearing that highlighted the approach of the year 2000 and its potential effect on computers. The authors and the news media began predicting that the year 2000 will bring on unprecedented catastrophe. But the year 2000 (Y2K) computer problem has, according to our analysis, been vastly exaggerated.

It also became apparent that authors were copying each other without really understanding the basic principles of the problem they were dealing with.

For example, one "documentary" demonstrated how Y2K will shut down the national power supply and paralyze the country. Obviously, the writers of the documentary did not understand the details and consequently replaced fact with fiction.

When analyzing the matter, we must let simple logic and common sense guide us. One reader clarifies Y2K in a very sober manner:

"[The] Y2K 'problem,' has been exaggerated by people who do not understand it, including those in the media.

"Pure logic suggests that the only affected operations are a function of the year, such as financial transactions that extend across the Millennium, such as mortgages and the like.

"Matters that are not a function of the year, but rather of the day or hour, will not be affected, as the examples below will show.

"Power generation and distribution are adjusted to

demand, which is a function of the time of year (not the year), weather, climate and the like. It would not be affected.

"Traffic signals are timed as a function of traffic flow and volume, so they are adjusted to operate at different phases during rush hour (high flow), night and weekends (lower flow), in other words, time of day and day of the week. The year does not enter into it.

"It follows that predictions that traffic will come to a standstill, that elevators will stop running and the lights will go out, etc., are pure nonsense."

—*Living in South Carolina*
published by Mid-Carolina Electric Cooperative, December 1998, p. 4.

During my early investigation of the matter, I consulted several computer experts, including my son, Joel, who is System Manager here at Midnight Call Ministry. We determined that the entire Y2K matter is mostly a "cloud of dust" that will blow away quickly.

For that reason, I determined not to get involved in the issue because "dust clouds" do come – but they disappear quickly depending on from where the wind blows.

However, as time went by, it became quite obvious that we could not sit still and let things take their course. When those who are computer illiterate began calling our office and, with trembling voices, asked for our advice regarding the approaching global catastrophe, we felt obligated to respond.

Such requests continually coming to our office let us know that something had to be done.

On the one hand, we would have liked to ignore the alarmists and the doomsayers. On the other hand, we saw great concern and confusion among many sincere people who wished to know the truth.

What would we tell an elderly lady who fearfully explained that taking money out of the bank and placing it under her pillow causes her to be more fearful and insecure? "What must I do?" she asked.

How would we answer others concerned about buying survival rations being offered by many organizations, including Christian ministries?

In short, we felt duty-bound to explain these developments and point out the errors of the alarmists, separating facts from fiction. That, fundamentally, is the reason we have written *When Y2K Dies*. ■

CHAPTER ONE

Ten Most Frequently Asked Questions

Ten Most Frequently Asked Questions about Y2K:

1. How dangerous do you think the Y2K virus is?
Answer: First, it isn't a virus but a simple computer software problem.

If your computer is "of age," say 10–20 years old, just reset the time as instructed in the operating manual and your computer should be functional.

A virus, on the other hand, is a malicious program specifically written to infiltrate computers by copying itself over and over again.

It carelessly or intentionally overwrites important data on the computer.

2. If Y2K shuts down my computer, will others be effected?
Answer: Your computer has no relationship with other computers unless you have your computer programmed to operate in conjunction with other computers which are

programmed to receive or dispense information to your computer.

3. If computers don't recognize the year 2000, will they shut down automatically?

Answer: Indeed, some computers may shut down because of software or hardware which doesn't recognize that "00" means the year 2000. The solution is described in answer 1.

4. Will the Y2K problem shut down our power supply?

Answer: First of all, power production doesn't exclusively rely on computers. Every power plant is operated from a control center and in that control room, there are wires, switches and innumerable other gadgets that are designed to be operated by hand.

5. What about water and gas supply? Will it be effective?

Answer: Again, neither water, gas or electricity is totally relying upon computer programs. Here too, a manual override is provided.

Furthermore, it is important to understand that the computers assisting the operation of these plants don't care what year it is, they are programmed for the seasons of the year and the hours of the day.

On a hot summer day, more water is needed than on a cold winter day.

Natural gas is needed more for heating in the winter than it is in the summer.

Obviously, the greatest volume of electricity, for example, is needed in the latter part of the afternoon when people return home from work and turn on the heating or air conditioning, use the shower and start cooking, etc.

6. One respected author claims that no electric production facility can guarantee that power will be available come January 1, 2000.

Answer: That is true. But neither can any power company guarantee that they will be able to produce power tomorrow either. This is a standard diplomatic answer from virtually all businesses, organizations or government branches that deal with the public. Can anyone in their right mind guarantee that an earthquake won't destroy a certain power facility? Of course not; therefore, your chance to have power supply come January 1, 2000, is just as good as your chance to have power tomorrow.

7. We have been told that traffic on land, sea and air will collapse due to the Y2K problem.

Answer: Again, computers which are programmed to assist in the operation of transportation don't relate to the year but, as already mentioned, they are programmed to regulate the flow of traffic during certain hours of the day. We must reiterate that transportation computers don't care whether it's 1999, 2000, 2001, 2002, etc.

Computers which control traffic lights, for example, have been programmed to accommodate the increased flow of traffic during rush hours, let's say 4–7 p.m. Streets, roads, and highways which are designed to carry a heavy load of traffic will need to have the traffic light programmed to a shorter "red period" and a longer "green period."

Quite naturally, electricity, gas and water must be made available during and after the rush hours in greater volume to cities and towns. The power consumption is at the peak between 5 and 6 p.m. on the east coast. At the same time, the west coast has excess power available because they are about 3 hours behind the east coast.

This fact again illustrates that Y2K has no direct effect on the power supply in the U.S.A.

8. Will banks shut down because the computer thinks it's 1900?

Answer: At this point in time, the computer can't think. It can only do what it is instructed to do. A car can't start by itself, move out of the garage and drive to work. The car is unable to do anything for itself; the driver operates the vehicle. The financial industry is very much computer reliant. Indeed, they must be Y2K compliant, and from all indications, virtually the entire industry is ready. If your credit-card expires in the year 2000 or beyond, then it proves that the bank is ready. This is also applicable to loans, mortgages, stocks, bonds, and other time-related financial matters.

9. If Y2K interrupts air traffic, will a large part of our economy come to a standstill?

Answer: The responsible organization for air traffic is the Federal Aviation Administration. They are way ahead in this field. For example, today, you can buy a ticket to fly from Los Angeles to New York on January 1, 2000 or virtually any date thereafter. This simply means the computer is programmed and Y2K is already dead.

10. I have heard that the world will be set back a hundred years at midnight December 31, 1999.

Answer: If that was the case, virtually all of the world's population wouldn't exist. This conclusion is based on the assumption that the computer can "think." That, however, isn't the case. Because of the vastly exaggerated media coverage calling attention to the potential problem of Y2K, computers the world over won't only be ready but for the first time in our "computer age," they will be well prepared.

CHAPTER TWO

The Next Millennium

Generally, people expect something extraordinary to occur when, at midnight, December 31, 1999, the world officially enters the third Millennium after the birth of Christ.

Great celebrations will take place around the world. For example, Vatican sources estimate that over 26 million people will visit Rome during the year 2000.

Throughout the land of Israel, new facilities are being erected to accommodate the expected flood of tourists for the 2000 celebration.

Members of every European nation are feverishly working on plans to make January 1, 2000 memorable.

Such preparations are also in full swing on the American continent. Even African and Asian nations cannot dismiss the significance of the Christian calendar which marks the year 2000.

Festivities have been planned for years. And one travel magazine reports that many of the world's favorite hotels are already sold out for that date.

In fact, *The Kansas City Star* reported as early as December 31, 1995:

"Still wondering where to go for this year's New Year's Eve party? Relax, you can squeeze in somewhere. Starting to make plans for Dec. 31, 1999? Forget it—you're probably already too late.

Even Mickey Mouse has run out of rooms.

Although it's still four years off, the changing of the annual odometer to 2000 has already shaped up as the biggest blast of the 20th century.

Guest lists are filled at some of the world's party hot spots.

The Rainbow Room in Manhattan: There are 470 persons ahead of you on the waiting list. The Savoy Hotel in London: The fortunate can enter a lottery for seats or rooms. Don't even try the Space Needle in Seattle: it's booked for a private party.

Reservations are piling up for the annual Kaiser Ball in Vienna∫at the La Tour d'Argent restaurant in Paris∫at the Waldorf-Astoria in Manhattan.

Looking for something a little more traditional? Colonial Williamsburg is full, and there are 107 names on the waiting list.

Walt Disney World in Orlando, Florida, reports all 17 company-owned inns are taken that night.

Why this mad rush?

'People look at New Year's as a time for a new beginning,' said Rainbow Room publicist Andrew Freedman. And this 'is really monumental in terms of new beginnings.'

Booking your table this far in advance also is an act of unrestrained optimism.

'Absolutely,' agreed Larry Wodarsky of Sacramento, California. 'The feeling was, we know where we want to be, and therefore we will be.'

Wodarsky, 49, president of the Money Store Investment Corporation, put down a $1,000 deposit to reserve Rainbow Room seats for himself and his wife; among the 198 others who did the same is an 87-year-old man.

A technical note:2001 is actually the first year of the new Millennium. But people are more excited about the calendar clicking from 1–9–9-9 to 2–0–0-0."

—*The Kansas City Star*, 12/31/95, p.A-4

Politicians are hoping that the new Millennium will usher in peace for the world. This hope is reinforced particularly by religious people, of whom the pope in Rome is the undisputed leader.

However, some disregard this date due to the claim that our calendar isn't accurate and that the actual year 2000 has already passed. Others say the year 2000 will arrive several years from now.

Hello, '96? Maybe It's Really 1999

Never mind that at midnight in Times Square it will already be early morning on New Year's Day in France, or that it will be only 9 p.m. New Year's eve in Seattle. There are far more profound discrepancies.

For one thing, the year that begins Monday probably should be numbered 1999 or 2000 or 2001. The problem has to do with a monk who lived a long time ago and meant well.

Dionysius Exiguus was asked by Pope John I in the sixth century to come up with a method for calculating the date of Easter that would resolve vexing differences that then existed.

The monk was an expert in astronomy and mathematics, and he came up with tables for fixing the date of Easter each year. While he was at it, he decided to renumber the years of the calendar to focus on the birth of Jesus instead of on the founding of Rome, the event that most of Europe used as a starting point.

Dionysius placed the birth of Jesus in the 753rd year of the old Roman calendar.

> Trouble is, Christian scholars generally agree that Herod
> the Great, king of Judea, died in the 750th year of the old
> Roman calendar—and the Gospels say that Jesus was born
> in Herod's reign.
>
> *—The Kansas City Star,* 12/31/95, p.A-1

Regardless of varying opinions, the Roman calendar is accepted globally as the authority and the entire world has programmed activities and schedules based on it. So what's all the fuss about the year 2000 anyway? Read on!

What is Y2K?
The letter "Y" stands for the "year"; the "2K" for 2 kilo, meaning 2000.

Besides those who expect great celebrations, other voices, primarily in the U.S.A., bid us to take caution. This claim is based on the assertion that computers will crash at the moment of the transition from 1999 to 2000.

This is because at the beginning of the computer revolution, programmers were advised to save memory by using two digits instead of four to represent the year. For example, they wrote 1973 as "73."

Thus, at the end of 1999, the last two digits of the year will go back to "00," causing the computer to assume it is 1900. Some say this will cause the entire program to crash.

Several issues of *Midnight Call* magazine have referred to the Y2K problem, prompting many of our readers to send us material about the potential problem.

Others have telephoned to express their great concern and have even accused us of negligence because we have failed to warn the public about the coming "global catastrophe."

We have collected material from multiple sources. Some analysts explain the matter in simple terms, confirming their apprehensions with facts and figures. Others base their claims on mere speculation and seem to aim toward creating uncertainty, even chaos, among their constituencies.

Many books have examined the possible scenarios that might unfold at the beginning of the year 2000, or even before.

To responsibly address this matter, we have decided, as mentioned in the introduction, to present the basics regarding the development of the Y2K problem and how it actually relates to the average private citizen, particularly Christians.

The Significance Of the Year 2000
First of all, we must point out that for Christians, no date, including the year 2000, has any particular prophetic message. When the Lord Jesus spoke to His disciples about end-time events, He clearly stated, *"Take heed that no man deceive you."*

Thus, the key word in the Lord's warning is "deception." What kind of deception is He speaking about? Answer: *"...Many shall come in my name saying I am Christ and shall deceive many."*

It is of monumental importance to understand these simple words of our Lord, who warns His disciples and in turn His entire church that the most terrible thing that can happen is deception, which means following someone who says he is Christ but is not.

Jesus does not warn us about the year 2000. He does not tell us to fill our storehouses with goods to prepare for a possible catastrophe.

The opposite is true; He clearly and emphatically instructs us in Matthew 6:31–34:

"Therefore take no thought, saying, What shall we eat? or, What shall we drink? or, Wherewithal shall we be clothed? (For after all these things do the Gentiles seek:) for

your heavenly Father knoweth that ye have need of all these things.

"But seek ye first the kingdom of God, and his right-eousness; and all these things shall be added unto you. Take therefore no thought for the morrow: for the morrow shall take thought for the things of itself. Sufficient unto the day is the evil thereof."

What Do Y2K Prognosticators Say?
Y2K prognosticators do exactly the opposite: They lead us to worry about the things of this world. They tell us to pre-pare for a catastrophe which they urge on with inflamed rhetoric.

However, such warnings are primarily flawed and dis-tract our aim and hope from Jesus. Y2K alarmists guide us toward the things of the world, and that, dear friends, is one of the great deceptions Jesus warns us about!

Why shouldn't we worry? Because we haven't received the spirit of fear and bondage but rather the spirit of liberty in Christ.

Comfort In the Midst Of the Uncertainty
Jesus comforts us: *"Let not your heart be troubled: ye believe in God, believe also in me. In my Father's house are many mansions: if it were not so, I would have told you. I go to prepare a place for you. And if I go and prepare a place for you, I will come again, and receive you unto myself; that where I am, there ye may be also"* (John 14:1–3).

We will do well to keep our focus in the right direction, which is expressed in our readiness at any moment to be in the presence of the Lord, regardless of any date or expected event which may cause panic among the children of this world. We shouldn't fear because our hearts and minds are anchored in the Eternal One! ■

CHAPTER THREE

Global Unity and Y2K

Well-respected authors often present a clear-cut case that the world is moving toward globalization, which will lead to the loss of sovereignty of individual nations.

While I agree with that interpretation, I would like to add that globalization is not something to look forward to in the future; it is a reality here and now.

Needless to say, any problem on the globe, whether it takes place next door or 10,000 miles away, affects us. Therefore, the Y2K problem serves to weld the world even closer together.

The few nations which have been reluctant to fully participate in an international computer language and global economy are now being forced to fall in step.

Governments the world over have issued guidelines to solve the Y2K problem. Private firms which are not up to date in fixing the Y2K problem will lose government contracts. This is equally true with all global manufacturing companies, trading firms, banks and organizations.

Executive Order 13073

Executive Order #13073–Year 2000 Conversion, issued by President Bill Clinton on February 4, 1998, includes the following orders:

• Cooperate in addressing Y2K problems with private-sector operators of critical national and local systems, including the banking and financial system, the telecommunications system, the public health system, the transportation system, and the electric power generation system.

• Communicate with foreign counterparts to raise awareness of and generate cooperative international agreements to address Y2K problems.

The attention surrounding the Y2K problem clearly solidifies humanity. Now for the first time, there is a problem which isn't limited to or confined within the borders of a country, but is truly global.

How each and every nation is interdependent has been demonstrated during the last decade, and has been particularly highlighted by the fall of the Soviet Union and the united effort of the Allies during the Gulf Conflict.

Although previous attempts at global unity have been dismissed as unworkable, global unity is now a requirement for peace and prosperity. Many experts claim that without serious progress toward global unity, the highly tensioned political climate could dissolve into a catastrophic world war.

I have dealt with global unity in my book, *How Democracy Will Elect The Antichrist.* The following excerpt demonstrates an important point:

"Today, the battles for new territory are no longer fought with weapons of war. The battles take place in political offices and corporate boardrooms, as well as through the media. It is now virtually impossible to take over a country by military force.

For example, consider Saddam Hussein's attempt to take over Kuwait. Virtually the entire world agreed to oppose

Hussein's aggressive move and defeated him militarily. Why did the world oppose him? Because his action was a threat to the efforts being made toward a peaceful New World Order.

Even assisting another nation without the popular support of the people has failed too, as was the case with America's war in Vietnam. Russia had a similar experience in Afghanistan. Shamefully defeated and embarrassed, they had to withdraw!

We are in a new world today. Different rules and regulations apply. After political unity is achieved, there won't be any need for diverse political opinions, or opposing parties. The people of the world will believe they have finally found the system that works to create a peaceful and prosperous world-wide society on Earth.

There will be no need for the overwhelming burden of keeping diverse governments, institutions, infrastructure in business, economy, commerce, industry, finance and military forces. God's Word says the people will "...have one mind" ("New World Order Rules," page 151).

This world-wide unity clearly prophesied in the Holy Scriptures, *"...these have one mind..."* is being established and solidified! ■

CHAPTER FOUR

A Realistic View Of Y2K

e have already mentioned a number of reports on the progress being made toward solving the Y2K problem before January, 2000.

All News Is Old News

When we quote a date of any publication, we must keep in mind that the actual information is much older. For example, the July issue of *Midnight Call* is prepared between the end of April and the beginning of May. So the news we publish in July is already several months old!

This practice, incidentally, is used throughout the publishing industry.

Furthermore, weekly news magazines are pre-dated. For example, the May 3rd, 1999 issue of TIME will arrive at our office about April 26, 1999.

So it stands to reason that gathering the news, putting it into electronic form, then printing it, takes time. To summarize, "All news is old news."

The Real News

I must not fail, however, to mention the exception to this maxim; namely, the Word of God. Some 6,000 years ago, God promised His salvation when He pronounced judgment upon the serpent:

"...I will put enmity between thee and the woman, and between thy seed and her seed; it shall bruise thy head, and thou shalt bruise his heel" (Genesis 3:15).

Approximately 4,000 years later, salvation came:

"For God so loved the world, that he gave his only begotten Son, that whosoever believeth in him should not perish, but have everlasting life" (John 3:16).

Jesus Will Come Again

This same Jesus, who came to give His life as a ransom for many, made this prophetic promise:

"Let not your heart be troubled: ye believe in God, believe also in me. In my Father's house are many mansions: if it were not so, I would have told you. I go to prepare a place for you. And if I go and prepare a place for you, I will come again, and receive you unto myself; that where I am, there ye may be also" (John 14:1–3).

The book you are reading now serves the purpose of pointing you to the eternal book, the Bible.

I want to make this clear: If there is any literature, including this book or our *Midnight Call* magazine, that hinders you from reading the Bible, get rid of it and open the Word of God. Therein, you will find the exclusive truths which are eternally settled in heaven!

All of our literature serves as a guidepost to the eternal truth found only in the Bible. With this fact in mind, let us continue on our subject "When Y2K Dies."

Travel Industry Ready For Y2K

Travel Weekly, a leading publication for travel agencies, published the following story in its October 5, 1998 issue:

> When the clock strikes midnight on December 31, 1999,
> guests will not be evicted from their rooms or asked to pay

hotel charges accumulated since the year 1900.

Although the Year 2000 (Y2K) problem is causing some worry in all sectors of society, if the hotel industry has its way, the date will pass without nightmares and business will go on as usual.

Because the element of time is so crucial to the travel business, industry suppliers have been working diligently on the potential problems, and the hotel industry has spent millions to develop and test new software and to create project management teams.

"Clearly, hotel companies are aware of the complexity and severity of the Y2K situation, and it has caused each of us to step back and assess our systems. There's been a great deal of activity," said Bob Bansifeld, assistant vice president of management information systems at Hyatt.

Hyatt started its Y2K project in 1995, and Bansifeld was confident his company will finish its Y2K tasks well ahead of time.

"We feel very good about the plans we have in place. January 1, 2000 will be another business day for us," said Bansifeld, who has headed Hyatt's Y2K team for the last year and a half.

"The bottom line is that customers won't be affected."

In tackling its Y2K problems, Hyatt began with core computer systems, such as reservations and operations.

Its technology team assessed the problems and made appropriate changes.

On the reservation side, the system is already in place: Customers can make reservations beyond January 1, 2000, and groups can make reservations well into the next century, according to Bansifeld.

Promus Hotel Corporation, developer of such brands as Doubletree Hotels and Embassy Suites, plans to finish with its testing by the end of 1998.

According to Lorna Brown-Raye, vice president of reservations at Promus, the company's core business systems, such as reservations and accounting, are scheduled to be completed this year.

She said Promus will soon test its connections with CRSs.

Of particular concern to the hotel industry is the switch technology that permits individual properties to communicate with the CRSs.

Thisco, a subsidiary of Dallas-based Pegasus Systems, has a bit of an advantage in debugging Y2K problems: Most of its systems were developed since 1989, when the company was formed, so its technologies have been Y2K compliant since creation.

Still, testing and perfecting the switch mechanisms is a massive project.

According to Bryan Donowho, vice president of Thisco, the company is creating a separate Y2K test system.

By having this Y2K-dedicated system, Thisco will be able to test its hardware (by manipulating dates and observing how the software responds) without interfering with current business.

Establishing interfaces with the CRSs is the next hurdle in the Y2K battle.

Thisco said these connections should be made by the end of October. Then the three groups involved—hotels, Thisco and CRS vendors—will run end-to-end verification tests.

Because the CRS allows agents to make reservations only 11 months in advance, the goal is to have these connections perfected by February.

In effect, that is the last minute for hotel companies. If they're not Y2K compliant at that time, agents won't be able to use the CRS to make reservations.

Agents, of course, could try somewhere else, and the non-compliant firm will lose out.

"The agent will have a lot of options, and if one chain has problems, agents will have a choice," said Donowho.

Other Pegasus products also will be Y2K compliant.

Note the words, "...the non-compliant firm will lose out." Again, we see money, the most important ingredient in today's world, is clearly a key factor in the Y2K problem. This is just another confirmation of the previously proposed theory that, after the year 2000, the world will see a much better business system based on computers.

Investment Firms

One of the largest and most well-known investment firms, Charles Schwab, has opened the Internet for investors so every person can do his own research, make his own decisions and buy and sell investments for himself. The company's publication, *The Schwab Investor*, deals, among other things, with the Year 2000:

Q: There's been a lot of talk recently about computer systems and the Year 2000. What is Schwab doing to prepare for the Year 2000?

A: At Schwab, we have dedicated substantial time, energy and resources to render our systems Year 2000 (Y2K) compliant, and we want to assure you that we're committed to maintaining and enhancing our levels of customer service through the year 2000.

We have a comprehensive plan in place to achieve our goal of enabling our systems to process data and transactions without material errors or interruptions as we move into the Year 2000. We currently have an experienced team of 225 managers, programmers and consultants working on the Year 2000 challenge. As of June 30, 1998, we had already completed modifying our critical trading systems to enable processing of Year 2000 data and transactions without material errors or interruptions (Vol. 2, Issue 5, Page 3).

Schwab is only one example, but it is quite obvious that no one wants to lose; everyone has targeted his investment portfolio to make money. The government, the banking industry and the business world are all working fervently to make each computer Y2K compliant.

It seems unrealistic to assume that the world's money makers are careless about this important matter and would miss the opportunity to amass great profit by letting the competition get the best of them. That, I repeat, is very unrealistic.

Federal Reserve Chairman Alan Greenspan Comments On Y2K During Q&A At His September 23, 1998 Congressional Testimony

Senator, let me say that we have spent a very considerable time on this, as you can well imagine. We have never had an episode like this previously, so it's not a question of adverting to some historical precedent and saying, "This is what is likely to occur as a consequence."

What we know is that a computer program is unforgiving. It cannot make a single mistake, or it will break down.

A number of programs were written 30 years ago, even 20 years ago—I wrote a lot of them myself.

I remember that I—because we had to conserve computer space and capacity, because it was so precious, that you would write your programs in an extremely compact way. And I don't remember having any significant documentation to the programs that I wrote. And it never entered our minds that those programs would still be around, with dropping the one (1) and the nine (9) in the calibration of years, as we got close to the Millennium. As a consequence, we don't know what's there. So we have to go all the way back and evaluate it—our individual programs.

What we have done at the Fed is to go back, program by program, test each block to see in fact when you introduce the zero-zero (00), whether the system broke down; to then start integrating, as we are now, our overall payment system evaluations with the individual banks in our system. And at this particular stage, I would say we're probably on track, that we can be fairly secure that the Federal Reserve System, that the domestic operation—is probably going to be okay.

But our systems are integrated with the rest of the world. And from what I can judge, even though major banks and the major countries are moving at a fairly rapid pace, we don't know, and may not know until the actual time arises, whether everybody is, as we call it, Y2K-compliant.

My suspicion is that we're going to run into a lot of problems. And as a consequence, we are doing a great deal of planning on what happens if it goes wrong.

And we are reasonably certain that concerns about whether the banking system is going to work is going to engender a significant increase in currency demand as we move in November and December of next year. And as a consequence of that, we have ordered a very major increase in the currency available.

I have no question that we're going to have very unusual things occurring. I think that concerns about "just in time" inventory systems not working will induce accumulation of inventories in the fourth quarter of next year which would otherwise [not] have occurred. So even if everything comes out exactly right, there will be a Y2K effect.

The following questionnaire has been circulated to evaluate the business practices of American companies and those companies they do business with.

The Scripture *"...these have one mind..."* takes on more meaning as we enter into the new Millennium!

Questionnaire To Assess Year 2000 Compliance v3.1

Use this questionnaire to assess progress toward Year 2000 compliance of your own organization and everyone you depend on including vendors, customers, bankers, borrowers, distributors, utilities, transportation, government, maintenance, security, etc.

Dear (Respondent):

The Year 2000 Problem with computers could have very serious consequences for all of us. To assess the risks, we all need more information about the problem. We must all monitor our own progress in fixing it as well as the progress of those we depend on.

Please respond to the following questionnaire as fully as possible and return to the interviewer below. Thank you.

When did the company take, or plan to take, the following steps toward fixing the Year 2000 Problem? (mm/yyyy)

	Initial Date	*Finish Date*
Awareness	1._____	2._____
Assessment	3._____	4._____
Repair/Replace	5._____	6._____
Testing/Validation	7._____	8._____
Implementation	9._____	10._____

11. If the company has finished assessment, then specify or estimate how many lines of computer software code must be checked for noncompliant date fields?

12. What percentage of this code has been checked already and repaired or replaced?

13. How many of the company's computer programs are mission-critical, i.e., necessary to remain in business or in an important line of business?

14. If the answer to 13 has increased over the past three months, what was the previous estimate?

15. What percentage of programs are mission-critical?

16. What percentage of mission-critical programs are not compliant currently?

17. What percentage of mission-critical programs have been repaired or replaced?

18. What percentage of mission-critical programs have been tested?

19. What percentage of mission-critical programs have been implemented?

20. What percentage of mission-critical programs are from third-party vendors?

21. What percentage of mission-critical programs from third-party vendors are at risk of noncompliance in January 2000?

22. Estimate the percentage of the company's programs that are not mission critical that will not be compliant in January 2000.

23. (Yes — No) Are some of the triaged programs that are included in 22 above possibly mission critical to one or

more of the company's customers or vendors?

24. (million dollars) What is the company's information technology budget for 1998 and 1999 combined?

25. What percentage of the company's information technology budget is allocated to fixing the Year 2000 Problem in 1998 and 1999 combined?

26. How much is the company spending on Year 2000 liability insurance in 1998 and 1999 combined?

27. How much is the company budgeting for Year 2000 litigation and liability?

28. (Years:) How long has the company's Chief Information Officer (CIO) worked as the company's CIO?

29. (Yes — No) Has the company lost key information technology staff in the past 12 months?

30. How many of the company's payroll employees are working on the Year 2000 Problem?

31. How many vendors are currently working on the problem for the company?

32. (Yes — No) Are there other major information technology projects underway at the company that are, or may interfere with achieving Year 2000 compliance?

33. (Yes — No) Does the company currently have access to staff resources to achieve Year 2000 compliance?

34. (Yes — No) Are there any vendors and customers who are important to the well-being of the company who might be seriously noncompliant in January 2000?

35. (Yes — No) Is the company preparing, or does it have contingency plans to assure the operation of the company in the event that one or more key vendors or customers fail in 2000?

36. (Yes — No) Is the company planning to acquire any key vendors or customers before 2000 because they can not achieve compliance in time?

37. (Yes — No) Is the company cooperating with others in the industry to establish Year 2000 standards?

38. (Yes — No) Is the company working with others in the industry to prepare contingency plans in the event of Year 2000 disruptions and failures?

39. (Yes — No) Have any government officials asked the company to cooperate with local, state, or federal contingency planning efforts in the event of Year 2000 disruptions and failures?   ■

CHAPTER FIVE

Y2K: Real People Voice Real Concerns

During the October 1998 Atlantic Coast Prophecy Conference, we posed the following question to our conference speakers: "Are you concerned about the year 2000?" Virtually all of the speakers revealed that they had little or no concern.

Some emphasized their opinion that the Y2K problem has been vastly exaggerated.

We also addressed this question to all participants after the last session of the conference.

I realized immediately that far too many had concerns about Y2K which time restrictions made impossible for us to hear out.

We rephrased the question and posed it only to experts in the field, such as computer programmers, technicians, scientists, etc., the largest group of people left at that point.

Then I asked each individual if he or she had scientific knowledge about the matter.

Only four candidates remained. Each had the opportunity to answer the participants' questions.

After these brief interviews, a number of spectators in the audience who had stayed behind volunteered to answer questions according to their understanding of the problem. Here are some examples:

From Ladylake, Florida

I am concerned about Y2K. I understand and can certainly relate to the fact that it is a domino effect because one thing affects something else.

I can prepare physically by having commodities on hand, and cash. One of the concerns I think about, though, are the elderly; I mean those who cannot prepare financially. I think of people who are in nursing homes. I trust in the Lord, but there are so many things for us to read and listen to and it would be nice to know a little more concrete what we should do.

Conference speaker: Dave Hunt

I am not personally concerned about Y2K and what will happen on January 1, 2000. The reason is because I can walk into my bank, ask them and they say,"It's O.K!" I can get an FDIC brochure in any bank. This document says there are four government agencies overseeing the banks regarding Y2K. They are taking care of it.

I have communicated with power companies; they too said they are compliant.

On the Internet, you can get in touch with the largest power companies in the country and again they will tell you that they are on top of this thing.

I don't think that Citibank or AT&T are going to lose a dime. These people are in it for the money.

To tell me that the shelves are going to be empty in the grocery store is a no-go with me. Computers have nothing to do with growing wheat or cherries and so forth. To think that Safeway trucks aren't going to be running because of some computer glitch and let their competitor get ahead of them is unrealistic. They aren't going to lose a dime.

Some say, "Well, even though all of the computers in the United States are O.K., what about all around the world?" My answer is that business people in other countries are just as smart as ours.

Others even claim you can get this virus which will jump from computer to computer. The truth is, it's not a virus, it's a programming problem and all they are communicating is data. I have talked to many experts, and I haven't found anybody that's upset about this thing. But it sells a lot of newsletters and books. It's alarmism in my opinion.

Now, there will be some problems. But traffic signals going down? What does a traffic signal care about what date it is? What about the imbedded chips? Most of them are in toasters, VCR's, ovens and so forth, and they don't care about the year either. I think that the Y2K problem has been overblown. There will be some problems here and there, but you have computer problems all the time. They'll solve them.

From Wilmington, North Carolina

Yes sir, I am concerned. The Bible tells us in Proverbs that a prudent man sees the danger and takes refuge. I believe that there is reason for concern. If we put our heads in the sand, where it may be more comfortable temporarily, we are going to miss a great opportunity to be a Christian witness in our community.

We need to know where we stand on such things as this so when the world looks at us we have an answer; we are ready. So I trust the Lord that whatever happens is going to be all right because I don't care if He comes back today before all of this breaks wide open. But if He does not, I believe that there will be a problem and I believe that we have the answer to that problem.

Conference speaker: Dr. Ed Hindson

I am not worried about the year 2000 at all. I don't think that the year 2000 has anything to do with Bible prophecy. There is nothing in Scripture that says Jesus has to come by the year 2000.

Because of the fact that the Y2K glitch is related to the year 2000, people artificially assume that 2000 must have something to do with the Second Coming of Christ. I personally don't think anything related to Y2K, or the year 2000, has anything deliberately or directly to do with the coming of Christ.

Jesus indicated that He would come at a time when nobody would expect Him, and since virtually everybody is expecting Him in the year 2000, I am going to go on record and say that I know Jesus won't come in the year 2000. Now, I know that He could, but I don't think He will because everybody is looking for Him.

I think that we have to remind people to balance out the facts of Bible prophecy with our speculations that arise from assumptions.

Many seem to preach the speculations as though they were the facts.

The bottom line is that Jesus could come at any moment. I do think that His coming is near, but that doesn't necessarily mean that His coming is here or that it has to be in the year 2000.

From Fountain Inn, South Carolina

I am concerned about the year 2000 but I don't panic at the thought of the year 2000 coming as far as the Y2K situation with the computers. I am concerned that the people who are concerned about the year 2000 are going to create more of a panic then the actual panic itself. The removal of money from the banks, the hoarding of some items, the wish to find generators, and in extreme cases, people actually selling their homes and moving to the country, I think will create more of an atmosphere of panic than is called for.

I think that for the year 2000 it would be wise to have hard copies of bank records, hard copies of Social Security records, things that the government as to do with so you are able to prove what resources you have in other ways than to just simply rely on computer records.

As far as a major disruption which could last for days or even weeks, I don't believe that we are going to see that type of disruption in services.

From Summerfield, Florida

Yes, I am concerned about the year 2000 problem with the computer systems and I think it's been not nearly played up as it should have been.

It should have been started years and years ago. We knew 2000 was coming quite a while ago.

I think that there will be a lot of small problems that people aren't thinking about.

The vast majority of the people really don't have the foggiest notion about what's going on.

My neighbor, for instance, says, "Well I don't have a computer." If you don't have a computer that doesn't mean that you aren't going to be [affected by] the problem of the year 2000.

If you have a Social Security number, you should be concerned. I think that we need to prepare ourselves, and trust the Lord.

We don't need to be dumb and blind and stupid; we need to be prepared.

There are lots of unusual confusions that are going to happen and we need to [have our] heads up and... get this thing fixed.

From Old Town, Florida

Yes, I am concerned with the year 2000. I believe there is going to be a problem… There was a CEO from Florida Power and Electric who came to one of our local churches and he stated that there was no way that they were going to have utilities after the year 2000. He said that they are going to shut down and for us to picture Florida black. So I believe that it is going to be a problem.

Conference speaker: David Webber

Yes, I am concerned, because I believe that this may very well precipitate the world into a state of confusion and may help to bring in a new system using marks and numbers and doing away with our present system. I see some real problems. The military and some business corporations say that they have solved the problem, but I really anticipate a lot of confusion, and a lot of trouble.

Certainly this could hasten us going into a cashless society that uses marks and numbers.

From Mt. Pleasant, South Carolina

I really don't like the word "concerned" because I have read the end of the book, and I know that it is going to work out and we are going to trust the Lord. But if I was going to use the word "concern," I think that the biggest thing that I would be concerned about is that believers in Jesus would miss this as an opportunity to greatly witness to those who do not know the Lord.

We can see people vacillating between panic and total disregard.

There are many opinions about both sides of that. I don't know how much damage is going to be done and I don't know how long some disruption is going to last.

Also, I am sure that we will experience mostly human

greed bringing problems into the situation.

But it is a wonderful opportunity for those who know and love Jesus to have extra things to share with people.

I don't truly think as believers that we have to worry about Armageddon; I mean the Lord has it in His hand.

We need to learn to not neglect or discount, but to look for opportunity for ministry in the midst of whatever the situation is and then to prepare wisely as we can, not just for the sole purpose of our survival, because really, we are better off dead than alive if we believe the Book.

However, we should prepare so that we will have opportunities to minister to those who are panicked. We can present the peace of the Lord to those who are without. We can give them the practical gift of provision that the Lord allows us to give in His name and let our light so shine.

So, no, I am not concerned; I am trusting in the Lord; I am looking at the situation and just asking, "Where would you like me to fit in, Lord, and what would you like me to do?"

From High Point, North Carolina

Well, I can say that I have felt directed to be prepared for the year 2000. I must add however, after living and working overseas and recognizing the tension that is on the rise because of causes outside of the Y2K situation, I have a greater concern just for the time before the year 2000.

I feel that we as God's family need to be prepared for major challenges that could result in the suffering beyond just finances and transition or postponement of career.

We need to be ready for something of greater degree and that is in no way to be discouraging or hopeless because we as God's family members know the end of the story, and that is my preparation as I think of Y2K.

I look at each day up until then as time for us to be

prepared for new types and new definitions of world leader-ship coming forward and adjusting, allowing the Creator to lead us, knowing that He is responsible for the final outcome for what is to occur.

Conference speaker: Dr. Thomas Ice

Well, I am not overly concerned about the Y2K. Obviously, there is something to it. But, because there have been people who are saying in essence that the world as we know it is going to come to an end, it's resulted in people spending bil-lions of dollars to try and fix it.

The main problem is said to be embedded chips, but some are saying only 10% of those chips pose a potential problem. I believe our industry can and will fix it.

In the United States at least, warnings have gone out quite well and virtually everyone concerned is preparing.

Within the Christian circle, we've got to realize who was the first to start talking about this. It was a guy named Gary North who is a post-millennialist. He saw this as something that he had been saying for the last 25 years. This is about the sixth or seventh crisis that he has predicted. None of the other crises have happened in which he believed that God was on the verge of judging Western civilization. North prophesied that our civilization would collapse socially, eco-nomically, and politically and that out of the rubble the rem-nant of the Christians would rise up to rebuild society resulting in a majority being converted to Jesus Christ; thus, the Millennium would come in before Christ returns.

I don't buy that as a Biblical view, but you can see why he would be taking advantage of this Y2K thing. He has taken the lead to say that it's going to be this major disaster. I think that a lot of the hype about it has come from guys like him.

Of course, we may spend a night or two in the dark.

The biggest threat, they are saying, in the United States is power plants.

But what I have read so far tells me that they are taking care of these things.

Great Witness Possibility?

The answers to our question, "Are You Concerned About Y2K?" made it apparent that most who expressed concern based their concern almost exclusively on what they had heard. Therefore, we must ask a number of logical questions regarding the preparedness of Christians.

If we are to follow the advice of those who tell us to be ready for Y2K by stocking food, taking money out of the bank, and equipping ourselves with survivalist gear, then we must ask the question, "How are we as Christians to behave ourselves in relationship to other people who have not prepared?"

Think for a moment: If I were to prepare by having an abundance of food and equipment in my house, then a catastrophe occurred, would I advertise to my neighborhood the food available at my place?

Let's assume I did that. No matter how much food I had stored, it would be gone in just a few days with my neighbors still standing in line at my front door.

Some say this will present a unique opportunity to witness for Christ. Again, we must ask, "How?"

During times of catastrophe, and this I can tell you from my own experience as one who grew up during World War II, most people are out for themselves.

If, for example, a food shortage is the problem, then millions of people will be roaming the countryside in search of food (providing, of course, they can drive their cars; filling stations won't be able to pump gas because the pumps operate on electricity).

Will People Receive the Gospel?

In the midst of such chaos, will people suddenly listen to the message of the Gospel and be converted?

Let's look at some Biblical examples: In the Book of Revelation, we read about the opening of the sixth seal, resulting in a great earthquake, the sun being darkened and the moon becoming as blood. How do people react?

"And the kings of the Earth, and the great men, and the rich men, and the chief captains, and the mighty men, and every bondman, and every free man, hid themselves in the dens and in the rocks of the mountains; And said to the mountains and rocks, Fall on us, and hide us from the face of him that sitteth on the throne, and from the wrath of the Lamb" (Revelation 6:15–16).

In chapter 9, when the angel blows the sixth trumpet, war breaks out and one-third of the population (possibly 2 billion people) is killed. How do the people react?

"And the rest of the men which were not killed by these plagues   repented not of the works of their hands, that they should not worship Devils, and idols of gold, and silver, and brass, and stone, and of wood: which neither can see, nor hear, nor walk: Neither repented they of their murders, nor of their sorceries, nor of their fornication, nor of their thefts" (verses 20–21).

Later in Revelation 16, we read that as a result, of the fourth angel pouring out the vial upon the sun, man suffers the greatest heat wave ever.

Do the people repent and seek the Lord?

"And men were scorched with great heat, and blasphemed the name of God, which hath power over these plagues: and they repented not to give him glory" (verse 9).

Proclaim the Gospel Today, Not When Y2K Comes

The Bible very plainly instructs us:

"Preach the word; be instant in season, out of season; reprove, rebuke, exhort with all longsuffering and doctrine. For the time will come when they will not endure sound doctrine; but after their own lusts shall they heap to themselves teachers, having itching ears; And they shall turn away their ears from he truth, and shall be turned unto fables" (2nd Timothy 4:2–4).

We are to diligently preach the Gospel to people everywhere, regardless of the year, month, week or day. Today is the day—not the day of a predicted catastrophe.   ■

CHAPTER SIX

How To Fix Y2K

Y2K has not only been recognized as a problem in the United States but world-wide as well. However, Y2K and its predicted problems have gained much more attention in the U.S.A. than other countries.

Europe
During my visit to Germany and Belgium in August 1998, I asked a number of people about their awareness of Y2K and virtually all answered, "It's not a problem and if it is, we'll take care of it."

Israel
During our Holy Land Tour in October of 1998, I inquired extensively about Y2K and how Israelis feel about it. Their answer was somewhat similar to that of the Europeans. In general, they feel the problem is well taken care of, because Israel is a high-tech country.

When we arrived in Jerusalem, our group was booked at the downtown Hilton Hotel. The Fall 1998 issue of *Hilton Magazine*, featured an article, "Israeli Ingenuity Debugs The Year 2000" by Jose Rosenfeld. Here are some excerpts:

"In a fight against the clock, Israeli software companies are offering successful solutions to overcome the Year 2000 computer bug which threatens to paralyze our technology-driven world at the dawn of the new Millennium.

It would be nice if we could just take January 1, 2000, off

the calendar. Why worry about grounded planes, large-scale electricity blackouts, malfunctioning life-saving equipment, inoperative communication links, inaccessible bank accounts and bungled financial transactions? But everyone from the President of the United States on down is warning all and sundry—deal with the Millennium bug now or else!

Richard Kearney, partner-in-charge of KPMG Peat Marwick's Global Year 2000 Consulting, explained that the bug 'is an unprecedented challenge not only because it reaches around the globe, but more so because it affects everyone at the same time.'

The recent Asian financial crisis demonstrated the tremendous interdependence of the global financial markets. Businesses and markets are linked in a complex technological web. For example, financial analysts today trade equities on all major exchanges throughout the world with the push of a button. Complex trades that call for the simultaneous buying on one exchange, say, New York, and selling on another, say, London, could experience execution failures if not properly debugged. In addition, the network of clearing firms, banks and other intermediaries that are heavily dependent on correct trade information may end up experiencing errors, Kearney warned.

Several Israeli firms, anticipating the crisis, began working on solutions in the mid-90s with great success.

One such company is Crystal Systems Solutions, Ltd., of Herzliya. A member of the Formula Group, one of Israel's largest software consortia, Crystal has attracted an impressive roster of clients. Among the best-known are Kraft Foods, Phillips Petroleum Co., Ford Motor Company, Pratt & Whitney, Blue Cross Blue Shield, Reynolds Metals, Ralston-Purina, the Farmers group insurance unit of Southwest Nominees Ltd., and other blue-chip firms.

Called C-Mill, Crystal's software automates the Year

2000 conversion process. C-Mill scans a company's system
for date-related fields and automatically converts most fields
to four-digit dates. What makes Crystal's technology attractive is that it covers a wide range of old computer programming languages, making it a one stop shop for companies
looking to convert all of their computer code.

A large part of Crystal's success can be attributed to part-
nerships with large consulting groups at home and abroad.
Most notably, Crystal serves as subcontractor to HCL
America Inc., and is teamed up with New York-based Ernst
& Young to jointly offer conversions in the US. In Israel,
Crystal, in partnership with ForSoft, has attracted the coun-
try's telecommunications conglomerate Bezeq, Bank Leumi
and United Mizrahi Bank.

Sapiens, based in Rehovot, has devised solutions for IBM
mainframe computers—the nerve center for most big organizations, such as insurance companies, banks, and retailer
chains with programs in Assembler language. These programs are the hardest to fix, and since some are system or
database exits, the potential damage if these programs are
not converted properly is huge. Sapiens' Falcon 2000
Workbench is an automated program that analyzes and fixes
the complex Assembler language programs to comply with
Year 2000 requirements.

Australia

At the end of December 1998, I flew to Australia, and again
took the initiative to find out Australian reaction to the Y2K
problem. Australians seemed slightly more concerned, obviously, because it is an English-speaking country where news
travels quickly because no translation is needed.

But my survey of reports in newspapers, magazines,
radio broadcasts and television programs, indicated that the
Y2K problem does not enjoy near as much attention as it

receives in the United States. This should not be surprising, for the U.S. is the undisputed leader in communication and, above all, entertainment.

News-related "entertainment" is well established. Two examples are enough to make my point: O.J. Simpson and President Clinton. Y2K has served as an attractive subject that allows the news media to add fantasy to its freedom.

Russia

News reports about Russia simply seem to say, in effect, "Don't worry. If the computer acts up, we'll turn it off, and then turn it back on again."

The Experts

After writing several chapters on the Y2K problem, I asked my son, Joel, if he had any information suitable for publication. He handed me the special issue of *Software Magazine*, subtitled "Year 2000 Survival Guide." This 64–page magazine deals exclusively with the Y2K problem. Inside, the cream of the crop, the brains of America, offer their services for companies who think they have a Y2K problem.

The inside cover of *Software Magazine* contains this advertisement by NeaMedia Technologies:

> "At last! A suite of modular, completely cross-platform, enterprise-wide, Year 2000 tools. 'ADAPT/Enterprise' is a comprehensive toolset that sweeps the entire enterprise, PC to mainframe, providing Year 2000 data and source code analysis, tracking and risk management, multi-platform COBOL data conversion and course code remediation, and proprietary COBOL migration to open, Internet-ready systems. In addition, ADAPT/Enterprise provides PC-based database, spreadsheet, EDI and comma delimited text file analysis—

all with Year 2000 project management support. If your Millennium headache is enterprise-wide, you need a powerful range of pain relievers: Adapt/Enterprise."

Page 2

Thinking Tool (Bridging the gap between business and technology) Where's the weak link in your Y2K and risk management effort?

Think 2000, a powerful business risk assessment and analysis software product, enables you to identify the weak links in your risk management process.

Think 2000 presents a visual map of your entire enterprise, reveals interdependencies, and simulates the business impact on your bottom line. Think 2000 is the only business risk management solution that:

• Generates a prioritized list of critical technology components

• Supports risk-based testing

• Streamlines planning and documentation for Y2K compliance

• Identifies risk of failure in supply chain

• Reveals weak links in Y2K and contingency planning efforts

• Provides documentation for due diligence, audit and litigation

• Simulates impact of alternative mitigation and contingency plans

• Allows import and export of enterprise and project data

Think 2000
    Business risk management designed to keep you in business. Contact us today for a free demonstration copy of
Think 2000.

Page 5: Princeton SOFTECH headlines its one-page ad:

"When You Fast Forward to the Year 2000...Make Sure
Nothing Important Gets Left Behind."

The ad states:

Will your applications work when the Year 2000 arrives?
Have they been tested in a realistic Year 2000 environment?
    Experts agree that you need to "fast forward" to the Year
2000. You must run stress tests against critical Year 2000
dates using a CPU date simulator and accurately aged test
data for realistic testing.
    HourGlass 2000 is the world's most widely used system
date simulator. It is the only date simulator that supports
every major 4GL, as well as most programming languages,
DBMss and TP monitors.
    Ager 2000 is the only date aging product to support
semantic date aging with Runtime Resolution and built-in
facilities for Target Dating, Global Date Advancement and
pivot-year Window Watching.

Page 7
    Platinum Technology offers help:
    So divide and conquer.
    There's a Y2K issue that's still up for grabs—Who's handling the time-consuming task of fixing 60 million PCs? Get

the end-user on your team, with Trans century Office.
Intranet or LAN deployed, it automatically guides users
through everything from correcting BIOS to fixing spread-
sheets, databases and applications. It's IT controlled, allowing you to monitor their progress. Most importantly, it treats
Y2K as a virus, safeguarding against new errors and freeing
your staff to focus on other mission-critical tasks—so you
come out a winner. Experience a free interactive demo of the
process infoWorld calls "simple and elegant."

Page 11
    Accelr8 Technology Corporation offers "Download
Ignition 2000"
    Accelr8 Technology's Free Year 2000 Indicator
    —The First Step in the Navig8 2000 Process for Mid-
Range Systems

Page 15 has this to offer:

Are you comfortable with the commitments you've made to
your whole company about Y2K compliance?
    Code audit is now mandatory. Recently, prominent indus-
try researchers have strongly recommended that all Y2K programs urgently include a secondary impact analysis phase
prior to code testing. The only way for you to confirm
exactly where you are in your Y2K program is to perform
this audit.
    This is not news to us. Primeon is in the business of pro-
viding Y2K audit and remediation services. In fact, we've
been checking and correcting the most complex applications
longer than any other service provider. Our hallmarks are
quality, speed and breadth of source language support—we
support over 30 languages and platforms. Over two-thirds of
our audit staff consists of computer scientists with advanced

degrees. This brain trust combined with our proven parsing tools and methods yields the most accurate service in the industry.

So, in order to get more comfortable with your Y2K position, contact the audit experts—Primeon.

Cap Gemini publishes a testimony from Blue Cross Blue Shield of Michigan:

Page 17

"By partnering with Cap Gemini America and utilizing their tools, methods, people, and expertise, the Blue Cross Blue Shield of Michigan Change of century Mainframe Applications project is projected to be complete months ahead of schedule.

"By utilizing the highly automated UDM for testing, we have proceeded at a faster rate than we would have without such a tool. Progress to date has been remarkable."

The ad continues:

If you can't say the same about your Y2K project, it's time to see what Cap Gemini America's Application Renovation Center can do for you.

Whether you need to review the status of your Y2K program; automate your Y2K renovation; check the quality of your renovated code; or conduct fast, efficient Y2K testing, Cap Gemini America has the toolsets and services to help you get the job done. That's a good reason to add us to your Y2K team.

Page 19

Programart Corporation asks the question: "Afraid your redeployed Year 2000 applications will drive customers away?"

They continue:

> You should be. Because once you've completed your Y2K
> redeployment, you may find your applications no longer sat-
> isfy service level commitments. In fact, those same applica-
> tions that you've modified for Year 2000 may now be
> undermining your ability to meet batch window and online
> response time requirements, turning your customers away
> and jeopardizing your business.
>
> But it doesn't have to be that way. With Programart's
> APM for Year 2000 package, featuring Strobe and
> APMPower, you'll be able to meet customer expectations
> and build customer loyalty. This unique, award-winning soft-
> ware enables you to identify and eliminate application per-
> formance defects that your reengineering efforts
> inadvertently introduce.

NETinventory proclaims:

> Page 21
>
> December 31, 1999:
>
> Champagne is iced up and ready to go but you're at work,
> grabbing a soda out of the vending machine while salvaging
> your company's PCs.
>
> Year 2000 problems in the PC Lan can result in applica-
> tion failures, wrong financial calculations, legal ramifica-
> tions and excessive user down time. Start working toward
> PC LAN Year 2000 solutions now. NETinventory, from
> BindView Development, automates inventory and performs
> analysis of your PC LANs. No more hiring temps to do man-
> ual searches. No more kicking employees off PCs for Y2K
> compliance checking.
>
> Find out where your PC-based Y2K exposures are, and
> solve them.

Micro Focus warns:

> Page 31
>
> "Think You're Ready For The Next Millennium? Maybe You Ought To Take A Second Look."
>
> Get 100,000 lines of code reviewed—FREE with Micro Focus Verify. Call for more information.
>
> Think your code is going to stand up to the year 2000? One way to gain confidence is to have your code checked by Verify, a new Y2K service offering from Micro Focus. Whether your code was fixed in-house or remediated offsite, Micro Focus will go on-site to check your remediated code providing you with reassurance as the countdown approaches.
>
> Micro Focus solutions can remediate 1,000,000 lines of code in as little as five days." We also provide you with the documented evidence of your company's due diligence in solving the Y2K problem.

The Source Recovery Company asks the following question in this half-page ad:

> Page 45
>
> "Missing Source?" Look no further.
>
> If you can't find it, recover it.
>
> Time is too short to consider anything else. the success of your Year 2000 project depends on one key element—the source code.
>
> SRC recovers COBOL and Assembler source code in the IBM mainframe environment.
>
> We can recover your code in far less time and at a much lower cost than it would take to rewrite it.
>
> If you have several versions of course, we can also find the right one for you.

Page 53

Optima Software, Inc. headlines its ad: "Watch Your Mainframe Dance."

Your mainframe runs the mission-critical apps you're pouring all those Y2K dollars into. It ought to really move when you need it to.

With Optima's Change Man software configuration management solution your IT infrastructure is more agile and responsive...a smooth-stepping partner that gives you:

• A direct route to automating software quality

• Better performance in application development

• More reliable code moving into production faster

• Process control and process consistency

You name the dance—"The euro Tango," "The Y2K Two-Step," "The M & A Merengue"—Change Man knows the moves. Get your mainframe dancing.

WRQ Reflection and Express Software offers this promotion:

Page 65 (inside back cover)

"Year 2000 problem on the desktop? Our company is completely on top of it." Even if your software vendors deliver compliant upgrades, you have to ensure that only those versions are used. You're also vulnerable to non-compliant PC BIOS chips, custom applications, and users who launch other non-compliant software. Failure can lead to erroneous mission critical data, misguided business decisions and exposure to liability.

Take charge now with the FREE WRQ Express 2000 Evaluator. Your free Express 2000 Evaluator software includes a fully functional, 30–day, 25–user version of WRQ's Express 2000 Suite.

It installs easily and enables you to start tackling the desktop problem right away:

• Remotely identify applications installed on PCs throughout the network

• Identify non-compliant applications and BIOS chips

• Prioritize upgrade efforts with accurate usage data

• Detect new applications and lock out non-compliant applications

Order your free Express 2000 Evaluator software today. Because if you're not covered at the desktop, the bottom could fall out of your entire Year 2000 strategy.

QES offers to solve the Y2K problem:

Page 66 (back cover)
Do Y2K tests manually or with scripting, and a team from Venus could win the World Series before you finish the job.

In Y2K conversions, testing is half the job. So if you're going to test manually or with scripting, you might get done in time for Y3K. But don't despair. There's QES.

We do testing without scripting. That's right. No scripting at all. It's all automated. In fact, it's so easy, anyone can do it.

But the best part is, you're not testing tests, you test actual business applications like invoices, paychecks and credit-cards. So the job gets done in a matter of weeks, not years.

Which means you save a bundle. Plus, QES provides you a legal record of your testing.

Experts Say: We Can Help!

These 14 firms which advertise in *Software Magazine* have one thing in common: They are saying "We can fix your problem!" Some of these companies guarantee that they can not only fix the problem but, should anything go wrong, they will take responsibility for it as well.

I took the time to read the entire magazine for a very specific purpose. I wanted to see if I could detect the same alarming voice which I hear almost daily from our readers and supporters of Midnight Call Ministry. I found no such concern. Also, I must point out that this magazine primarily deals with the negative aspects of Y2K; otherwise, it would contradict the firms offering their services to fix the Y2K problem.

Advertisers pay for the magazine; they are the source of the publisher's profit. As in virtually the entire media industry, any type of media must appeal to the possible largest segment of the public so the media can tell businesses, "Look, we have access to 100,000; 1,000,000; 10,000,000; 100,000,000 people."

Subsequently, they determine the cost for the business to advertise in the publication. Weekly news magazines such as *Time, Newsweek, U.S. News and World Report*, and monthly publications such as *National Geographic, Reader's Digest*, etc., are primarily geared to the general public, while publications such as *Hunter's Guide* and *Medical Journal* are specifically directed to one segment of the population.

Obviously, because *Software Magazine* is concerned solely with computer professionals, they cannot belittle the danger of Y2K. In fact, they must emphasize the urgency of bringing all computers into Y2K compliance since that's the source of much of their profit.

Although the publication covered the entire Y2K scenario well, I found no evidence to support the panicked reports coming from the survivalist camps and many Christian ministries.

In fact, of the 66 pages in the magazine, only half of a page dealt with the survivalist theme. Here is the entire article:

Make Mine Freeze-Dried

Ready to head for the hills around Christmas 1999? Wired magazine has already profiled some IT types who are. Perhaps it's with folks like them in mind that Y2K News Magazine, a small community newspaper, hosts advertisements such as these:

- "If Noah were alive today, he'd be building a water tank" is the tagline for watertanks.com, a company that sells tanks, water barrels and more.

- "Millennium gourmet food reserves. With Y2K less than 500 days away, we must have food storage!" That's an 800 number that boasts longest guarantees, fastest shipping, not to mention maximum nutrition and shelflife.

- "Emergency candle burns 100–plus hours. Case of 12 burns three hours per night for a year. With S+H and holder $68.15, U.S.A. made since 1970. Compare and burn your money wisely!" There's a Web site ready to give you more detail.

- "Free firewood, mulch, even compost. Easy, innovative ideas to prepare …your family for emergency readiness."

That from a Texas P.O. Box.

Of course, the paper, which is aimed at consumers, local community leaders, church groups, and so on, offers plenty of lively feature material and any amount of classified ads from Y2K consulting services and tools providers too. But if you're looking to joining the ranks of the freeze-dried faithful, it's not a bad place to start.

—Software Magazine, 10/15/98, p.13

We do not need to belabor this point to show that *Software Magazine*, a professional source, does not see Y2K as a threat worthy of "survivalist" action. As a matter of fact, this feature is apparently mocking the survivalist industry because *Software Magazine* only quotes some advertising from various suppliers, but fails to give actual telephone numbers or addresses.

On the same page, we read about Appleton Paper, a $1.3 billion company, which had fixed their entire system by February 1998:

Y2K Reading

All Done Five Months Early

At every Y2K forum, speciality paper maker Appleton Papers gets plenty of kudos from the date-change doyens for having taken plenty of action early. In fact, the $1.3 billion Wisconsin company—which makes the carbonless copy papers used in multi-part forms—claims to have certified 5.5 million lines of code as compliant last February.

Appleton's story is told in a recent issue of Compuware InTelligence, an in-house publication for IT executives published by test software vendor Compuware Corporation, Farmington Hills, Mich. To get hold of a copy, E-mail: Intelligence@compuware.com or call (248)737–7300.

—Software Magazine, 10/15/98, p.12

I asked my son, Joel, who deals with programming on a daily basis, to find another industry publication which prominently features the Y2K issue. As of this writing, he has found none.

Again, it is not our attempt to arbitrarily simplify the Y2K problem or belittle its significance; neither are we aiming to whitewash the potential problem, because, obviously, a non-compliant computer system can create problems.

But we do wish to expose the exaggeration of the message heralded by reputable scholars and ministries.   ■

CHAPTER SEVEN

Y2K Subject From the Air

This chapter of our coverage on the Y2K issue doesn't originate as usual from my office in West Columbia, South Carolina.

As I write this, I am a passenger with the Olive Tour group on my way to the land of Israel.

El Al Flight 0010

El Al Israeli Airlines Flight 0010 is positioned 10,100 meters (34,000 feet) above the Atlantic Ocean. We are cruising along at about 1,050 kilometers (653 MPH) and the outside temperature is -52 C (-63 F).

The captain has just switched on the "Fasten Seat Belt" sign due to expected turbulence.

Seat belt fastened, and having just consumed a delicious ribeye steak, I am ready to think about the now-infamous and often frightfully presented subject titled "Y2K."

Fear Not

Our position on Y2K is solidly built on the Rock of our salvation, the Lord Jesus Christ. He promised to never leave or forsake us and He stated very clearly, *"I will build my church"* (Matthew 16:18).

Our Lord, who died for us that we may have eternal life, actually instructed us not to worry. We can repeatedly read words of comfort, such as *"Fear not"; "Be not dismayed"; "Cast all your burden upon Him";* and *"He careth for you."*

To His disciples and others gathered around Him, Jesus made this most assuring statement, *"Therefore I say unto you, Take no thought for your life, what ye shall eat, or what ye shall drink; nor yet for your body, what ye shall put on. Is not the life more than meat, and the body than raiment?"* (Matthew 6:25).

Although our Lord draws a grim picture of the endtimes by mentioning famines, pestilences, wars, rumors of wars, earthquakes, etc., He never told us that we should spend our thoughts or time being alarmed about things we cannot change.

Jesus not only said that we should *"...take no thought, saying, What shall we eat? or, What shall we drink? or, Wherewithal shall we be clothed?"* (Matthew 6:31), but He added something remarkable, *"But seek ye first the kingdom of God, and his righteousness; and all these things shall be added unto you"* (verse 33).

If we would only heed our Lord's invitation to seek the things that belong to the kingdom first, we could avoid many heartaches, confusion and even tragedies. He is in charge and will take care of us; He already has won the battle for us! With these wonderful, encouraging, and uplifting words of our Lord in mind, let's take another look at the "Y2K" problem, this time from my seat on an El-Al Israeli Airlines Boeing 747 Jumbo Jet en route to the land of Israel.

Computers

We all know that computers have become part of our modern society. Virtually every person, particularly in the Western World, depends on computers in more ways than one. Computers organize our paperwork, and help us with our bills, taxes, income, expenditures, and much more. In offices, they are indispensable. Actually, computers govern much of our private lives. We start and end our day with computers. Our alarm clocks that wake us each morning are probably operated by a computer chip and our coffee machine kicks on only after a pre-programmed computer engages the electric current. The electrical system that starts our cars and directs virtually all of its other functions of the vehicle is probably operated by a computer.

After a day's work, we still depend on a number of computers. Even the traffic lights that stop us on the way home are operated by computers.

We could fill the entire book with examples clearly showing that we are a computer-dependent society.

Computer Needs Man

What is a computer? In just a few words, a computer is an electricity-driven apparatus made of metal, plastic and other components. Its primary purpose is to receive and dispense information. It receives and dispenses information via programs written by computer programmers.

The computer consists of two main components:

1) hardware, the physical computer; and
2) software, the programs that tell the computer what, when and how to manipulate data.

The computer cannot perform functions outside the realm of instructions given to it by the programmer. But, unlike humans, the computer retains all information it is fed.

As an illustration, think of a student who instantly and fully understands everything his teacher presents. We'll call such a student a "Wunderkind."

Even the most excellent student reaches his maximum mental capacity generally after 7–10 years of university study leading to a Ph.D. After that, the student supposedly knows virtually everything about the subject and should be able to practice what he has learned.

Let's compare that "Wunderkind" with a computer. As already mentioned, the computer retains all instructions it receives. But that's only the beginning.

While a student normally receives instruction from several college professors for 4 – 10 years, the computer can receive and keep information indefinitely.

The student's brain capacity is definitely limited, in contrast to the computer's "brain," which is expandable, therefore unlimited. In practical terms, the computer can retain all the knowledge available in the world.

Computer Versus Man

Let's analyze further the capacity and capability of a computer in comparison to man.

Could the smartest person in the world remember the names and address of about 100 people?

Probably.

But would it be asking too much to add each person's birthdate, Social Security number and telephone number? I believe this person has reached his limits.

If such a person, indeed, exists, he would be an extreme rarity. But say he does.

Now could he add even more information to his bank of knowledge about those 100 names, such as their passport number, driver's license number, bank account numbers, utility account numbers, home mortgage account number, car loan numbers, and insurance policy number? I think it's clear that no man on Earth can retain such a tremendous volume of information.

Yet, computers do it easily, instantaneously, and totally accurately.

Program Problem

But there is one little problem: Despite its seemingly infinite memory capacity, the computer cannot think for itself.

Subsequently, "wrong" information entered into a computer could prompt a string of unintended computer function, possibly causing it to shut down.

Year 2000 Examples

Here at the office, my son Joel Froese tested our computer system by changing its operating date to 12/31/99. When the clock turned to 12:01 a.m., January 1, 2000, he found that each of our office computers recognized "00" as a date in the year 2000.

On a recent tour to Europe, I asked our European representative to disconnect his computer, remove all batteries, then replace, reconnect and restart the computer after changing the time and date to 11:59 p.m., December 31, 1999. We waited one minute and the computer recognized "00" to be in the year 2000, although this computer was over ten years old.

In the meantime, I have collected a great volume of material from sources around the world and, in most cases, computers that have been tested have revealed no difficulties in recognizing the year 2000.

In another case, I asked one of our board members, Jerry Brown, who is responsible for the computer network for the South Carolina Department of Licensing, about this matter. He stated that after testing virtually all of the computers under his jurisdiction, only one in 10 did not recognize the year 2000. He said he subsequently fixed the computers which seemed to malfunction relating to the date.

Of course, that doesn't mean that many computer systems might not recognize the year 2000. There is no need to belabor the point that such an event could cause havoc.

At this time, we are counting the days we have left to fix and test each computer, including their operating systems.

These few illustrations show that the computer problem can be fixed, is being fixed, and must be fixed. Those who are experiencing difficulties, or are computer illiterate, can take advantage of hundreds, if not thousands, of firms who are offering their services to fix potential Y2K problems in their systems.   ■

CHAPTER EIGHT

The New World Of Peace

T he scenario being proposed by a number of respectable and well-informed authors presume the Y2K-engendered catastrophe will initiate the ascent of a world dictator. He will, they say, appear on the horizon at the moment of utter chaos and offer a solution to the global crisis. The world will greatly admire this person to the point of worship. What's our opinion about such a prediction?

While the Bible makes no mention of the year 2000, it does speak of global takeover, global catastrophe and global unity that will occur during the time period called the Great Tribulation. Most serious eschatologists agree that the world will become one based on four aspects:

1) Political   2) Economic   3) Military   4) Religious

To better understand this coming global world and its relationship to Y2K, let's first understand that, on one hand, the Bible concerns itself primarily with the nation of Israel, while on the other hand, all other nations are called Gentiles. There is no in-between. Christians must recognize this truth about Israel.

This, incidentally, is also true on an individual level. We are either saved or we are lost. There is no in-between.

John 3:36 makes this clear, *"He that believeth on the Son hath everlasting life: and he that believeth not the Son shall not see life; but the wrath of God abideth on him."* Either we have eternal life or we don't.

Nationalism

Unfortunately, Christians often permit themselves to be brainwashed by nationalism, which is sometimes cleverly camouflaged as *patriotism*. It is natural, of course, for every nation to think it is the greatest. This tendency can be traced to Lucifer, who wanted to be equal with God. Thus, nations in general think more of themselves than others do.

For example, on one hand, for Americans there is just simply no question and no debate: America is number one, the greatest, the best, the strongest.

Canada, on the other hand, is not so bold in its proclamation but Canadians would not, under any circumstances, change and become Americans because Canada, for the Canadians, is the greatest thing on Earth.

A strong sense of nationality undergirds Australians, as well. For example, just recently, when I was in Australia, I mentioned to a number of people there that I had resided in their country for five years. As if pre-programmed, each one immediately asked, "Have you regretted leaving Australia?" For Australians, Australia is the greatest nation on Earth and they couldn't imagine anyone wanting to leave. This type of self-glorification is natural and is found in virtually every nation of the world.

Paul the Hebrew

The faithful servant of the Lord, the Apostle Paul, was quite different. He actually had all the reasons in the world to be proud because he was, by birth, a member of God's chosen race. In Philippians 3:5, he writes, *"Circumcised the eighth day, of the stock of Israel, of the tribe of Benjamin, an Hebrew of the Hebrews; as touching the law, a Pharisee."*

If anyone could be proud of his heritage, Paul indeed could. But what does he say about his national, tribal and family heritage? *"I...do count them but dung that I may win Christ"* (verse 8).

It is not surprising, therefore, that the Apostle Paul received such an abundance of revelations from the Lord to pass on to His church. He was empty of himself. There was no national or family pride in his heart.

In chapter 2:3 he admonishes, *"Let nothing be done through strife or vainglory; but in lowliness of mind let each esteem other better than themselves."* This surely is an uncomfortable word for modern psychology-based Christianity, which tries very hard to instill self-esteem to each individual.

These points may not seem to relate to Y2K, but they must be emphasized because if our hearts are humbled before our Creator, and we have no pride left, then we can begin to understand His plans and begin to clearly see His intentions for those of us who love Him.

The Last World Empire

To reinforce the fact about Israel and the Gentiles, permit me to list some of the aspects that identify the last world empire. The Prophet Daniel, in captivity in Babylon, received visions about the entire Gentile world in relationship to his country, Israel.

In Daniel chapter 2, the four-fold power structure of the Gentile world empires is shown to us, beginning with Nebuchadnezzar as the first empire, and ending with the last, the Roman Empire.

Chapter 3 documents Nebuchadnezzar's attempts to unite the entire world through religion by erecting a 60 x 6 cubit gold statue in an effort to force all the people of the world to worship the image.

It is significant that he uses music as a tool with which to unify the people and prepare their minds to worship the image:

"Thou, O king, hast made a decree, that every man that shall hear the sound of the cornet, flute, harp, sackbut, psaltery, and dulcimer, and all kinds of music, shall fall down and worship the golden image" (Daniel 3:10).

It is also remarkable that six musical instruments prepare the atmosphere for "worship" of the golden image.

Chapter 4 shows how the first world ruler is led to recognize that the God of Israel is the Lord of this world, *"Now I Nebuchadnezzar praise and extol and honour the King of heaven, all whose works are truth, and his ways judgment: and those who walk in pride he is able to abase"* (Daniel 4:37).

Chapter 5 documents the end of the first world empire, Babylon, and the beginning of the second world empire, Medo Persia.

Chapter 6 demonstrates Daniel's unwavering faith in the God of Israel which rescued him from the lion's mouth.

Chapter 7 reveals the deeper mystery of the fourth world empire, emphasizing the diversity of the Roman Empire.

Chapter 8 offers more details regarding the climax of this last empire which produces a leader identified as *"...a king of fierce countenance,"* the one we call the Antichrist.

Chapter 9 focuses on the sin of Israel, with which Daniel identifies himself, and the coming of the Messiah, ending in the breaking of the covenant by the false messiah.

Chapter 10 gives Daniel great courage coming directly from God, who calls him *"O man greatly beloved...."*

Chapter 11 reveals the world today and the spirit of the Antichrist, as it relates to Israel.

Chapter 12 gives personal comfort to the greatly beloved Daniel and his people, Israel.

These twelve chapters in the Book of Daniel provide a concise description of the entire history of mankind, revealing the rulership of the Gentile world powers in opposition to the power of God, who will establish world dominion through His people, Israel.

Two Powers

It is important for us to understand that from God's perspective, there are only two powers: His and the Devil's.

The Bible clearly tells us that the Devil is the prince of darkness, the god of this world, and he rules all nations.

I must add, however, that the Devil's rulership is subject to God's permission. So when we speak about the Y2K problem, we must keep in mind that God is always in control. He permits the Devil to build his kingdom on Earth, using any and all tools that are in and of the world—including political philosophies, whether democracy or communism, Nazism or dictatorship, socialism or monarchy. All systems fundamentally oppose God's system.

Theocracy

God's system is theocracy. For that reason, God chose Israel and prepared it to be a nation that would rule the world, based on His righteousness.

But the Creator of heaven and Earth, in His foreknowledge, knew that Israel would not be capable of ruling the world in truth and righteousness unless they were righteous. Thus, God gave them the law by which Israel should live. We all know they failed miserably.

That, however, never annulled God's intention of bringing about a peaceful society based on His principles. The fact that Israel failed and broke God's covenant does not change His eternal resolution. He will yet establish His Kingdom on Earth, just as He originally planned!

The Coming King

The Prophet Isaiah proclaims, *"But with righteousness shall he judge the poor, and reprove with equity for the meek of the Earth: and he shall smite the Earth with the rod of his mouth, and with the breath of his lips shall he slay the wicked.*

"And righteousness shall be the girdle of his loins, and faithfulness the girdle of his reins.

"The wolf also shall dwell with the lamb, and the leopard shall lie down with the kid; and the calf and the young lion and the fatling together; and a little child shall lead them.

"And the cow and the bear shall feed; their young ones shall lie down together: and the lion shall eat straw like the ox.

"And the sucking child shall play on the hole of the asp, and the weaned child shall put his hand on the cockatrice's den.

"They shall not hurt nor destroy in all my holy mountain: for the Earth shall be full of the knowledge of the Lord, as the waters cover the sea.

"And in that day there shall be a root of Jesse, which shall stand for an ensign of the people; to it shall the Gentiles seek: and his rest shall be glorious.

"And it shall come to pass in that day, that the Lord shall set his hand again the second time to recover the remnant of his people, which shall be left, from Assyria, and from Egypt, and from Pathros, and from Cush, and from Elam, and from Shinar, and from Hamath, and from the islands of the sea.

"And he shall set up an ensign for the nations, and shall assemble the outcasts of Israel, and gather together the

dispersed of Judah from the four corners of the Earth" (Isaiah 11:4 – 12).

This clearly promises a real kingdom of peace that will be established on planet Earth. True righteousness will be practiced on Earth because the righteous One, the Prince of Peace will then rule. Not only will the people on Earth be different, the animal world will be changed as well: *"...the lion shall eat straw like the ox."*

This is bad news for those who proclaim that they will, in their own strength under the auspices of Christianity, establish the true kingdom of peace on Earth. It can't happen. Only God can do it.

Preparation For Peace
Preparation for this kingdom of peace is now in full swing. The outcasts of Israel, the Jews, are returning from the four corners of the world to the land of Israel. This regathering has been taking place for over 50 years and will climax in the gathering of all the Jews from the world's five continents to the place God has prepared: the land of Israel.

It is important to realize the Lord God will bring about the regathering of the Jews, and He is the one who will change the heart of man.

Devil Cast Into Bottomless Pit
Revelation chapter 20 describes how this will take place, *"And I saw an angel come down from heaven, having the key of the bottomless pit and a great chain in his hand*

"And he laid hold on the dragon, that old serpent, which is the Devil, and Satan, and bound him a thousand years,

"And cast him into the bottomless pit, and shut him up, and set a seal upon him, that he should deceive the nations no more, till the thousand years should be fulfilled: and after that he must be loosed a little season.

"And I saw thrones, and they sat upon them, and judgment was given unto them: and I saw the souls of them that were beheaded for the witness of Jesus, and for the word of God, and which had not worshipped the beast, neither his image, neither had received his mark upon their foreheads, or in their hands; and they lived and reigned with Christ a thousand years.

"But the rest of the dead lived not again until the thousand years were finished. This is the first resurrection.

"Blessed and holy is he that hath part in the first resurrection: on such the second death hath no power, but they shall be priests of God and of Christ, and shall reign with him a thousand years" (verses 1–6).

Thus, we see the cause for the world's chaos is Satan, the great deceiver. Only when he is locked up in the bottomless pit will the world experience real peace.

How does this relate to Y2K? The answer is rather simple. Y2K is just another stepping stone toward Satan's establishment of his own kingdom of peace on Earth. He can use Y2K's unifying power to assist his plan.

However, his kingdom will not last a thousand years, but only seven, the time period described in prophetic Scripture as the Great Tribulation.

Keep in mind that there will be no peace outside the peace which the Lord Jesus has established. His peace is based on the price He paid; namely, His shed blood! ∎

CHAPTER NINE

The Coming World Leader

The Book of Daniel indicates that the final Gentile empire will be ruled by one person, the Antichrist. Under his leadership, this new, unprecedented successful global system will prosper as none before.

Note that the leader of this global empire will *"come in peaceably."* He will receive power *"by flatteries"* because *"he shall work deceitfully."*

If that's not politics based on our much-cherished Roman democracy, I don't know what is! No politician who speaks of wars and destruction, and who promises suffering and hardship, will be elected. Today's politicians cannot afford to tell the truth; they must resort to lies to gain the favor of the people.

How About Doubling Taxes?
One of the most pressing burdens in the United States is the national debt. Any politician knows that the national debt can only be paid for by taxes. Let's assume a candidate for a political office came on the scene and told the people:

> "Fellow Americans, we are in a dire need. Our expenditures continue to outperform our income.
>
> "Sooner or later, we will be in debt so deeply that we can't pay the interest on the money we borrow each year to

operate this great country.

"I, therefore, propose that for the next five years, we double all of our taxes and add $5 per gallon gasoline tax. After five years, this will erase all of our debt.

"Our cash reserves could then monopolize the rest of the globe with the accumulated budget surplus."

Would this candidate be elected? Never! To draw a majority vote, he would have to tell his constituencies that taxes will be *cut*, not raised.

For some strange reason, the voters believe the lies rather than the truth. For the past 80 years, taxes have steadily increased in spite of political promises to reduce them.

Such political debates are mentioned in Daniel 11:27, *"...they shall speak lies at one table...."* These are undeniable Biblical facts!

Globalism Is In

Globalism is in; nationalism is on its way out. Soon or later, more and more nations will have to sacrifice some of their national sovereignty. Prosperity is increasing and has reached a height never before dreamed of.

Yet we know that destruction will come. The Great Tribulation, the Bible tells us, will be the most horrendous time the Earth has ever experienced. The Lord Jesus stated, *"For then shall be Great Tribulation, such as was not since the beginning of the world to this time, no, nor ever shall be"* (Matthew 24:21).

Peace Or Catastrophe?

Which comes first? Global peace and prosperity or global catastrophe?

Based on my understanding of the Scripture, I believe that great peace and prosperity on a global level will precede the Great Tribulation. Never before has the world (particularly the Western World) enjoyed so much peace and prosperity as today. To reinforce this fact, note the following article from a British publication:

"The world will enter 2000 in very much better shape than our predecessors predicted. Fatalism is out of fashion, argues Madsen Pirie.

"It is because people invest numbers with magic that certain dates acquire momentous significance. At various times in our history, sections of the human race have expected the world to end, or, at the very least, for doom and disaster to unfold. Some thought at the turn of the previous Millennium that one thousand years was the right date for the Second Coming.

"There have been other dates since, when faithful followers have sat on hillsides waiting for Armageddon, only to disperse disappointed until their leaders could recalculate the date."

"As the calendar turns once more, there have indeed been wars and rumors of war, and earthquakes in diverse places. Surprisingly, though, there has been no widespread "millennial movement" predicting the world's end.

"Indeed, large sections of humanity seem to be facing the new Millennium with a degree of optimism that would have been alien to our forefathers.

"What is missing is the sense of fatalism, the feeling that human beings are helpless objects to be tossed by fate like corks upon the waves. In its place is the widespread conviction that humanity is, to a great degree, master of its own fate, and that what happens to us will be the result of human action, even purposeful action.

"The doomsayers have found little echo to their apocalyptic alarms.

"The environmentalists have been confounded by evidence that so-called 'scarce resources' are falling in price, indicating that supplies are outstripping demand.

"Most of the raw materials which were supposedly dwindling show greater reserves now than they did two decades ago. It seems that our ability to conserve and to employ substitutes is better than expected.

"Similarly, the talk of man poisoning the habitat has been difficult to reconcile with water and air that are becoming cleaner year by year, and with productive processes that visibly pollute less than their predecessors did a century ago.

"Those who spent the century predicting the final demise of liberal capitalism were confounded when the system of collectivist central planning imploded instead.

"Even those who thought the British economy was impossibly outmoded and doomed to defeat at the hands of the "miracle" economies of Germany and Asia, have been confounded by a flexibility in the old-fashioned free-market model which has left it quite able to stand up against its rivals.

"Other portents of doom have included a breakdown of social values and a so-called collapse of civil society. After decades of relentless gloom, the family remains the most popular of institutions. Most people settle down at some point with a partner of the opposite sex and raise children.

"A recent poll of the 16–21 age group showed that "to be happily married with children" was the most popular aspiration, coming out ahead of a successful business career.

"The threat of the developed world groaning under the burden of an aging population has been dispelled. People do indeed live longer, but their active lives are also prolonged, and timely pension reforms have ensured that most of the elderly in the early decades of the new century will be capital rich, not requiring support from taxpayers.

"As the new Millennium comes upon us, it seems that even the four horsemen have lost their thrall over us.

"Appalling famines still occur in localized conditions as a result, of human folly, but Malthus has been confounded by a world food supply which has been increasing faster than the population which it has to support.

"War still occurs, also on a local scale, but no longer threatens the planet each day with nuclear annihilation. Pestilence is still there, but smallpox has been conquered, and polio subdued.

"Few doubt that the early years of the new Millennium will bring victory over malaria, AIDS, and many forms of cancer. Although the fourth horseman, death, is till with us, advances in diet and medicine mean that each year he takes longer to catch us."

—*The Economist*, Special Insert, "The World In 1999" (p. 19)
by Dr. Madsen Pirie, president of the Adam Smith Institute

The Coming Peace

My understanding of the prophetic Word indicates that it is essential that the world prosper in peace and harmony in order for the Antichrist and his system to find global acceptance.

You cannot deceive people to the point that they will worship the Antichrist by oppressing them or by creating a time during which chaos rules.

Prosperity opens up horizons whereas poverty leads to regression.

With democracy now in firm control, we are witnessing steady progress toward a unified global political system, in spite of many setbacks.

From an economic and financial point of view, the world is already interdependent, marching to the steady drumbeat of global unity.

Interdependency
Who would've thought that the stock exchange crash in little Hong Kong could initiate such a devastating effect on the entire continent of Asia, with the shock waves registered throughout the Western World? This clearly teaches that globalism will result in great prosperity, but it will also bring great dangers for the national sovereignty of individual countries.

No longer can any nation, even the United States of America, afford to act independently of other nations. Isolated nations such as North Korea, Cuba, and, lately, Iraq clearly illustrate that it has become impossible to operate successfully outside the global family.

For these and many other reasons, I am convinced that peace and prosperity will precede the time of the Great Tribulation.

When They Shall Say "Peace And Safety"
When the Apostle Paul writes to the Thessalonians about the Rapture, he highlights one sign relating to the world, *"For when they shall say, Peace and safety; then sudden destruction cometh upon them, as travail upon a woman with child; and they shall not escape"* (1st Thessalonians 5:3).

Notice that this is addressed to *"they"* and *"them."* Paul makes the clear distinction in the next three verses when he identifies the church. Note here the words *"ye," "we,"* and *"us."*

"But ye, brethren, are not in darkness, that that day should overtake you as a thief. Ye are all the children of light, and the children of the day: we are not of the night, nor of darkness. Therefore let us not sleep, as do others; but let us watch and be sober" (1st Thessalonians 5:4–6).

Realizing that today's world is more educated than ever, we have no reason to believe that the people who say "peace and safety" are just *imagining* "peace and safety." Here again, "peace and safety" precedes "sudden destruction."

That, I believe, presents a powerful argument against the possibility of a world-wide chaos predicted by those who believe that Y2K will usher in the Antichrist. Based on the Scripture, we will have peace first and then destruction!

Light Stronger Than Darkness

Bible students know that the events of the Great Tribulation preceded by peace and prosperity will be initiated by the appearance of the Antichrist.

However, the Antichrist cannot make his appearance until the Light of the world has gone out. Keep in mind, he is the product of the prince of darkness.

It's impossible for darkness to work parallel to the light. The church, according to the words of Jesus, is *"...the Light of the world."*

The only way this light can be extinguished is by taking the church out of this world via the "Rapture."

Therefore, we may expect the Rapture to take place at any moment!

Scripture doesn't clearly reveal the time of the Rapture. Rather, the Bible warns that it will occur *"when ye think not,"* and, therefore, admonishes us to *"Watch ye therefore, and pray always, that ye may be accounted worthy to escape all these things that shall come to pass, and to stand before the Son of man"* (Luke 21:36).

Don't Wait For Y2K!

Another reason we reject the argument that Y2K will usher in the Apocalypse is that we are repeatedly warned to wait and prepare for the coming of the Lord; we are never admonished to wait for the Great Tribulation, the Antichrist, or the year 2000.

Of course, we must be watchful always, but we need not be alarmed, especially since this predicted scenario is primarily based on speculation.

Our Lord wants us to be ready for His coming; there is no better way to be ready for the Rapture than being busy in His work.

Jesus said, *"Blessed are those servants, whom the Lord when he cometh shall find watching: verily I say unto you, that he shall gird himself, and make them to sit down to meat, and will come forth and serve them"* (Luke 12:37). ■

CHAPTER TEN

Facts Or Fiction?

The Y2K problem does exist. Those who ignore this fact might pay some unpleasant consequences. But it is clearly unwise to assume that the world will come to a standstill because of a Y2K-engendered crisis.

Farmers will continue to plant and harvest their fields. Cattle will graze on the pasture; chickens will lay their eggs, and dairy cows will produce milk. Potatoes and corn will continue to grow. All of these natural processes will continue despite the Y2K "problem."

Scripture guarantees this: *"While the earth remaineth, seedtime and harvest, and cold and heat, and summer and winter, and day and night shall not cease"* (Genesis 8:22).

Furthermore, agricultural processing plants will not stop preparing and packaging food to ship to grocery stores. Truckers will "keep on trucking ... on the road again," transporting that food from plants to grocery stores.

Someone may say: "But tractors and virtually all machines are computer operated and, therefore, they won't work." Computers can be disconnected and virtually any truck or equipment can operate without them.

Business As Usual
On occasion, during wintertime in South Carolina, we might get some snow (in the amount of a few flakes that one could almost count). When such a snowfall threatens, the news media exaggerate the coming "snow storm" to such an extent that the night before, people hurry to the grocery

stores to empty the shelves of milk and bread. Such "news" is good for business.

This example shows that in our affluent world, most people are well-educated and they know how to prepare. But they usually overdo their preparations by far.

Businesses are working extra hard so they won't lose a nickel.

Americans, with a labor force that works harder than in any other country in the world besides Japan, will no doubt be in business making money January 1, January 2, January 3, etc.

Incidentally, computer down-time isn't unusual. It happens each day the world over. When the computers at a grocery store crash, the manager simply has the cashiers ring up sales using calculators. The manager doesn't sit down, put his head in his hands, and say, "Let's quit! The world has come to an end!"

Mind Working Overtime

To demonstrate how the public in general and Christians in particular have been guided in the wrong direction is quite evident from one letter I have selected:

Dear Brother Froese:

Many favorable words have been written about the benefits of computers, with almost none in opposition.

In just 9 months, the computer world will enter year 2000 with few computers equipped for the digital change.

Most people, including myself, have asked, "What's the big deal?"

I have since learned the possible consequences of being indifferent to this seemingly simple problem and have become convinced that this minor "glitch" could be the instrument that could give the 666 man of Revelation control of the world with the blessings of a grateful populace. It could even force a peace treaty between Israel and the Arabs.

For example: This one-world commercial system that man has created is controlled by computer chips high above the Earth or on the ocean floor. Since computers must work in unison with each other, mal-equipped ones will either not function at all or else give false information. Imagine the chaos of a world that has no computers to operate its power plants!

No electricity means no water or heat for homes, no pumps to pump gasoline and oil for transportation, no telephones, no mail service, no stock market; banks and hospitals will be paralyzed. Panic would reign supreme.

The one that controls electricity in today's society controls the world. This would be the most opportune time in history for the Antichrist to emerge with a solution and become man's master.

An eager and willing world would gladly accept him as God and his 666 mark in order to survive.

Please give some remarks on this possible scenario.

—A Christian Brother

The entrance of the Antichrist scenario this letter-writer proposes is unrealistic because Antichrist will *"come in peaceably"* and *"by flatteries"* because he *"shall work deceitfully"* (Daniel 11:21,23). ◼

CHAPTER ELEVEN

Y2K Benefits

L et us now look at the potential benefits of Y2K from a layman's perspective.

Each one of us has either moved from one home to another or even cleaned out a garage, basement, attic, or storage building. Remember how much "junk" you had to dispose of?

Even after holding a garage sale to get rid of some items, you probably still had leftovers: boxes of outdated magazines, old tennis rackets, plastic toys, a mountain of old clothes and piles of paraphernalia from your old school.

Obviously, when you stored these items, you intended to use them sometime. Well, plenty of time had passed but you had not found use for any of it.

Even the memories you had wanted to hang on to had become things of the past. Most people agree that getting rid of the accumulation of items somehow makes us feel better. It frees up more space, cleans up the place, and often instills us with a desire to take on another home improvement.

With that in mind, do we need to be told how much a computer clean-up would improve productivity?

Computer Space
"Space," the storage capacity of a computer, was at a premium just ten years ago. Each "byte" was precious.

The first computer we bought for our office in 1980 was a CPT Word Processor, an amazing machine that retailed for

$20,000. The computer came equipped with two disk slots, each of which held a large, 8–inch disk that could retain over 400,000 bytes. Nine such disks could hold the entire Bible. This was a tremendous improvement over the old card file and mechanical flexowriters operated by a hole-punched paper tape which we had previously used. Those early machines are useless today.

From "Kilobytes" to "Gigabytes"

Computer memories have become larger and faster while their physical size has become much smaller. Also, the good news is that the price is continually coming down. Storage space, now counted in gigabytes, is virtually unlimited.

But we still have to be careful in managing the accessible workable memory, otherwise, it can become useless. A poorly organized computer filing system is like a room full of books that are randomly stacked.

We would have to look through hundreds of books in order to find a specific bit of information. But a well-organized filing system is comparable to a carefully catalogued library in which we can retrieve specific information efficiently and quickly.

The Good News About Y2K

Old programs were rather cumbersome and often caused more pain than gain. I would not want to guess how many of today's computers are working at only 5–10% of their capacity because of outdated programs.

We can all begin to understand the great advantage a thorough computer "house cleaning" could create.

In fact, Y2K may not be a problem after all; it might serve us a great benefit, as "Ajax" does for a dirty bathroom.

Note the following article, which appeared in *Software Magazine*:

Bad Taste, Good Medicine

"'So quit with the long faces about your Year 2000 project. It's good for you.' So says RCG Information Technology, an Edison, N.J.-based IT consulting firm that has come up with this Top 10 list of benefits:

1. *A Competitive Edge.* A company that has achieved Year 2000 compliance before its competitors will be able to redivert resources back to company growth initiatives sooner. That could translate into market-share gains.

2. *Cheaper Insurance.* With a strong Y2K compliance program, business risk insurance should be easier and less expensive to acquire. Many insurers are requiring a Y2K plan or proof of compliance before issuing new policies, and banks are eyeing lines of credit in a similar light.

3. *Faster Time To Market.* A strong Year 2000 plan will help companies effectively evaluate their suppliers and help ensure that supplies are available so business can proceed as usual.

4. *Clearer Big Picture.* CIOs rarely have time to stand back and evaluate their current software infrastructure. The Year 2000 conversion process allows an organized and structured examination of the infrastructure and the opportunity to weigh what works and what doesn't.

5. *Better Documentation.* With the evaluation of all systems through the Year 2000 compliance process, levels of documentation will be improved.

6. *Getting Rid of Dead Wood.* Systems that the Year 2000 compliance process shows no longer provide effective

functionality or that provide redundant information can be eliminated. This will significantly reduce maintenance resource requirements.

7. *Cleaner Systems.* As a process converts line by line, Year 2000 compliance projects will help identify inactive code and provide the opportunity to clean up libraries, so freeing up disk space.

8. *Improved Testing Process.* A structured Year 2000 program will involve a testing and quality assurance package that can be used for testing of future apps.

9. *Easier Maintenance.* Older system infrastructures often contain a myriad of date routines. Through year 2000 conversion, a company can migrate to a standard date routine, which makes maintenance requirements easier to address.

10. *More Detailed Mapping.* Year 2000 conversions seek out all locations where a date field is used in a company's system. Other fields can be identified and a map of system-fields-created knowledge that can be invaluable when updating, say, account numbers that may exist in several places.

—*Software Magazine* (Special Issue)
"Y2K Benefits," Oct. 15, 1998, p.12

■

CHAPTER TWELVE

Conspiracy, Deception, and Y2K

Conspiracy has a number of different definitions, which vary according to the context in which it is used. The two definitions Webster gives for "conspiracy" are:

1) The act of conspiring together.

2) An agreement among conspirators.

The word "conspire" also has a two-fold definition:

1) To join in a secret agreement to do an unlawful or wrongful act or an act which becomes unlawful as a result, of the secret agreement;

2) To act in harmony toward a common end: ("Circumstances conspired to defeat his efforts").

American's don't use the term "conspiracy" to describe our Founding Father's proclamation of independence for the United States of America. But British literature outlines that historical event in terms of a *conspiracy* headed by George Washington and his associates. The British viewed this as an unlawful and wrongful act that opposed the prevailing authority of the Crown.

Positive Conspiracy

Let's call the establishment of the United States of America a conspiracy, but a positive one, in that our Founding Fathers "conspired" to become independent of Britain so they could form a new nation on the North American continent.

In the first half of the 1700s, Britain had established a number of relatively prosperous colonies on the East coast of North America. This area was desirable real estate because of its similar climate to Europe.

Colonies in the southern hemisphere such as Africa, Australia, Asia and South America, could not attain the prosperity of North America. This was because southern colonies primarily existed on the harvesting of raw materials and agricultural products which fetched a lucrative price in Europe.

Europeans, particularly with a good education and involved in business, and craftsmen, had little interest in leaving the continent for the unknown and dangerous tropics. Many who did, did not fare too well.

North America was different. It was the "New Europe," the new world.

Europeans could not only feel comfortable here, but after a relatively short time, they established themselves in trade, industry and agriculture equal to or better than that in Europe.

In short, America became the only country that superseded Europe. But again we have to keep in mind that this country was founded on a conspiracy—albeit a positive one—which in this case was staged against the prevailing government of Britain.

Another Positive Conspiracy

During the Second World War, a number of conspiracies were planned against Hitler, a man who had so much power that he not only controlled Germany, Austria and other European countries, but who was also the major source of the terror that came upon all of Europe. In general, we don't call such courageous undertakings a conspiracy, but that's what they were. Therefore, we must differentiate between positive and negative conspiracies. Most of us would agree that this case was a positive conspiracy.

Power Of Deception

Hitler's power, it is important to note, was not based on force; rather, it was primarily drawn from deception.

Some time ago, a discussion on National Public Radio revealed that at least 46% of German physicians were active members of the National Socialist Workers Party (Nazi Party).

Physicians are a segment of the population which, in general, have the best education. They surely wouldn't be counted among the masses of people who are easily persuaded by cheap propaganda. Yet, a large percentage of physicians decided to become purveyors of Hitler's ideas. Fully aware of who Hitler was and what he was proclaiming, they must have recognized his political philosophy and suspected where it would lead. Nonetheless, they chose to become members of the Nazi Party.

Royal Deception

Germany, as we know it today, and as it was under Hitler's rule, has been a relatively new country. Previously, it

existed as a number of independent cities and a confedera-
tion of kingdoms ruled by royalty, primarily from the house
of Habsburg in the south and Prussia in the north. In the
German language, the royal families and their descendants
are usually identified with the word "von" in front of their
family name.

In the upper echelon of Hitler's amazing fighting
machine, innumerable leaders were of royal descent.

One of the most famous was Werner von Braun. With his
help, America successfully competed with the Soviet Union
in the space race.

In fact, in the early 60's, an Australian newspaper
reported von Braun's arrival in America with the headline,
"U.S. Welcomes Hitler's Rocket Master."

Werner von Braun had been only one of many who
eagerly served in Hitler's fighting machine.

Other well-known names of great German rank were:
Werner von Blomberg, Feder von Bock, Carl von
Clausewitz, Karl Freiherr von Eberstein, Wolf Graf von
Helldorf, Gunther von Kluge, Adolf Viktor von Koerber,
Lutz Graf Schwerin von Krosigk, Erich von Manstein,
Freiherr von Richthofen, and Baldur von Schirach (from
Adolf Hitler—Das Ende der Führer-Legende.)

Many members of royal descent actually surrendered
their royal titles in order to identify with the average
German people and thereby honor Hitler.

Support for the system and the government was overwhelming.

That was the real reason Germany held out against virtually the entire world for six long years.

Deception To The End
Not only did the deception cover Germany like a thick blan-
ket of fog but even the enemies of the system, those whose

lives were threatened, were deceived as well. *The Jerusalem Post International Edition* reported:

> Betty Scholem wrote to her son Gerhard (Gershom) about everything—from the state of her bank account and her passing colds, to the unrest in Germany after the resignation of Kaiser Wilhelm and, later, the iniquities of the Nazis.
>
> Some 900 of her letters have been preserved in the papers of Prof. Gershom Scholem, the world-renowned expert on Kabbala. Nearly a third of these are now available in Hebrew in *Mikra Gershom Shalom V'IMo*, published by Schocken.
>
> Besides shedding light on the personal background of the man who founded the academic study of Jewish mysticism, they provide a touching portrait of a German-Jewish woman who slowly came to realize that her beloved homeland had betrayed her.
>
> The letters span approximately 30 years, from the end of World War I to the aftermath of World War II. Betty Scholem had a compunction to write, and wrote well.
>
> Artur Scholem, head of the family, owned a prestigious printing press which did a lot of work for the government. He expected all members of his family to be patriotic Germans. He was horrified when one of his four sons became a communist; another, a pacifist, was sentenced to a seven-month jail term for participating in an anti-Kaiser demonstration.
>
> To make matters worse, his boy Gerhard showed an undue interest in Judaism. Disgusted, Artur Scholem ordered his son never to cross the threshold of the family home again. He was civil enough, however, to congratulate Gerhard on the completion of his university studies, though he did wonder what a man with an interest in mysticism would do for a living.
>
> A terrible war was raging west and east of Berlin, but

Betty's letters hardly mention it. Only toward the end of 1918 did the war begin to be felt in Charlottenburg.

"We've all lost weight, though we eat well. There is not enough oil. I'm fed up. It is also extremely cold. The carpet in the big room looks like a dust rag. I could not wait for peace and normal life, and I bought a new one. Very nice, very expensive, too," wrote Betty in March 1918.

However, 10 days before the end of the war, she notes with satisfaction the end of censorship and the fact that an "extremist social-democrat" of her acquaintance had been appointed deputy minister of state. Eventually, she too became involved in politics and joined the Liberal Democratic Party.

In 1923, Gershom Scholem emigrated to Palestine, joined the staff of the Hebrew University, and married Hamburg-born Else Burchard. When asked how his son would manage to keep a family on a scant academic salary, the elder Scholem, who died of pneumonia in February 1926, replied: "It's his worry."

Betty visited Gershom in Palestine twice, but refused to move. Instead she went to live with another son, Erich, in Berlin. Life in Germany was changing rapidly—for the worse—but Betty did not refer to the changes.

"With us, political upheavals cause, in the first place, uncertainty in business. Unfortunately, we have had our lesson in the matter. The moment Hitler came to power the business suddenly came to a halt...Hitler never stops making speeches on the radio, in which he does not say much. But it is a fact that newspapers are closed and republican officials are dismissed on every level....For the moment the Jews have no grounds for worry," she wrote on February 20, 1933.

A week later, however, she excitedly reported that her son Werner, a former communist member of the Reichstag, had been arrested for attempted arson. "It can't be serious, for he

had been ousted from the communist Party and in the past
few years devoted all his time to studies. Possibly it's nothing but election propaganda."

How wrong she was—Werner was killed in a concentration camp.

Just a month after Werner's arrest, Betty writes: "The
political change is proceeding more quietly than on any former occasion. We must wait and stay silent. I really hope he
will prevent disorder." A week later she reports that Jewish
doctors had been dismissed from hospitals, adding hopefully:
"But one expects they will be allowed to continue in private
practice."

Gershom Scholem, situated in Jerusalem, was much better
informed than his mother about Germany. He writes her: "As
regards rumors, which you say are 90% lies, our opinion is
quite different from yours. Abroad, one is conscious of the
fact that one can destroy the Jews without violence."

And indeed Betty Scholem finally realized herself that
this time something was truly different from what she had
seen before. "Don't there exist 10,000 decent Christians—
well, 1,000 decent Christians—who would refuse to take
part in this process, and voice their protest? What has happened to the lawyers who, from one day to the next, lost all
income, may occur tomorrow to physicians. Probably this
cannot happen very soon to merchants, for Christian suppliers would not like to forgo the money of their Jewish customers."

People asked her why she did not go to Palestine, where
her son lived. She wrote him: "I hope that the possibility to
flee will always remain open to me. I am totally relaxed. But
I cannot be unmoved about the fate of my family here. Am I
right?

—*Jerusalem Post International Edition*, "Thy Mystics Mother"
December 14, 1998, page 22

What a story! It is almost unbelievable that a Jewish person who heard the voice of Hitler, who saw the soldiers march through the street openly humiliating Jews and destroying their businesses, could not bring herself to believe what was happening. That is powerful deception!

Strangely enough, though, all the conspiracies against Hitler and his Nazi empire failed miserably. He was able to hold on to power until about 60 million people had died. Over six million Jewish people were systematically murdered. Immeasurable suffering was inflicted upon Europe and much of the world, and Germany's cities were devastated, with millions of civilians, women and children killed. No conspiracy was designed that could end this mad undertaking of one diabolically inspired man.

Why did this happen? Because the nation was deceived. Therefore, the essence of conspiracy is not the conspiracy itself; rather, it is the one word, "deception." ∎

CHAPTER THIRTEEN

Global Conspiracies

Many articles and books by well-meaning, respectable authors claim that virtually all international movements, organizations, and societies are involved in a *global conspiracy*.

Some of their most outlandish statements focus on the founding of the United Nations and other global organizations. Such statements really muddy the water. My research has proven that most of these allegations are primarily the product of overactive imaginations.

Poor and Rich Countries

Obviously, there are some legitimate reasons to be concerned about a global conspiracy. For example, rich countries are worried about poorer countries. The rich ones try to create ways and means for developing countries to prosper. There is no advantage in having poor neighbors. The poor are not good customers. So naturally, rich countries try to "conspire" to make poor countries able to become paying customers.

In return, poor countries often try desperately to "conspire" against the rich so they, themselves, can become prosperous.

To assume, however, that these poor nations, usually known as Third-World countries, will receive the power through the United Nations or any other organization to force the rich nations to "share the wealth" is presumptuous.

International Purpose

There may be conspiracies within the movement of the United Nations among certain groups of nations, peoples or branches, but the United Nations, as an international organization, has a very defined purpose. The body of the U.N., among other things, legislates international laws, agreements, and treaties, and if necessary, enforces them.

Some may say, "That's exactly what we don't want or need." I disagree. We do need international cooperation governed by international law.

Let me list some examples. If any country drops a nuclear bomb into the ocean, it concerns all nations, particularly those who are closest.

If Mexico, for example, experimented with nuclear devices or if Canada experimented with biological weapons, the U.S. would most certainly be greatly concerned.

Hitler's Nazi machine proved the need for international cooperation among the nations regarding peace and world stability. Therefore, to call the United Nations a useless organization that only wants to stifle America's freedom is off-base.

Subject To Government

As Christians, we should always be willing to hear, accept, and proclaim the truth. Unfortunately, much untruth, tainted by nationalism, is being propagated throughout American Christianity in our day. We must learn to separate the truth from the lies.

Christians have not been delegated to change the political reality of this world. Whether we like it or not, agree or disagree, the U.N. Charter and the laws of the United

Nations are a matter of fact. If the Constitution of the U.S. opposes the international laws, then sooner or later the international laws will win. Global government will come sooner or later, whether we like it or not.

The Bible plainly teaches that we should be obedient to our prevailing government, *"Let every soul be subject unto the higher powers. For there is no power but of God: the powers that be are ordained of God. Whosoever therefore resisteth the power, resisteth the ordinance of God: and they that resist shall receive to themselves damnation.*

"For rulers are not a terror to good works, but to the evil. Wilt thou then not be afraid of the power? do that which is good, and thou shalt have praise of the same: For he is the minister of God to thee for good.

"But if thou do that which is evil, be afraid; for he beareth not the sword in vain; for he is the minister of God, a revenger to execute wrath upon him that doeth evil" (Romans 13:1–4).

Holy Nation

We must never forget that, as born-again Christians, we belong to a holy nation which is almost 2,000 years old. This "holy nation" is not of this world, *"But ye are a chosen generation, a royal priesthood, an holy nation, a peculiar people; that ye should shew forth the praises of him who hath called you out of darkness into his marvellous light"* (1st Peter 2:9). One day, however, the order will go out from Jerusalem superseding all other laws and regulations of every country in the world. We are serving the Lord Jesus Christ, who bought us with His own precious blood when He died at Calvary's cross and cried out, *"It is finished!"*

Wherever we may live, we are subject to eternal laws. But while on Earth in our perishable "tabernacle," we are subject to the law of the land.

Internationalism

Later, we will identify the greatest conspiracy of all time and
show that the present development of all nations—regardless of whether they are capitalist, communist, nationalist,
or a dictatorship—is part of the greatest conspiracy in history.

Whether it's the United Nations, the Council on Foreign
Relations, The Club Of Rome, the Vatican, or the
Bilderbergers, each has a distinct purpose with a defined
goal of furthering international cooperation so that we can
live together in peace and harmony. But this peace and harmony will not be a lasting one, and will ultimately lead to
the greatest catastrophe the world has ever known.

"Top Secret"

We know that the Bilderbergers usually meet privately in a
secluded place. This has caused many to speculate whether
indeed the mysterious group has something to hide.

Some time ago, someone sent me several pages of text
regarding the Bilderbergers which was marked "Top
Secret."

After reading the material carefully and doing some further research, I found none of the accusations justified label-
ing the Bilderbergers as an organization involved in
national or international conspiracy.

Every organization certainly has the right to limit its
membership and, subsequently, select who may attend its
meetings.

For example, I am not a deacon in my church. When the
deacons meet, I am not permitted to attend their meetings.
This is true with virtually all organizations and businesses.
Each has its own specific agenda and its meetings are not
necessarily open to the public; we certainly cannot call such
meetings a "conspiracy."

Furthermore, the article the reader sent regarding the Bilderbergers was obviously not "top secret." If it was "top secret," the sender would not have been able to obtain the material. Only those with the proper identification, access codes, and authorization would be able to receive such information.

Whenever we run across other so-called "top secret" information, we should take that designation with a grain of salt. If you and I are reading it, it is definitely *not* top secret.

Freemasons

We have read and heard about an alleged Freemason conspiracy; many claim they are supposed to rule the world. Investigating such allegations, however, has not convinced me.

They may be a "secret organization" with excellent international connections whose members try to assist one another in business and politics, but those characteristics don't qualify the Freemason Lodge to be labeled a conspiratorial organization.

The End-Time Puzzle

How does this discussion of conspiracy relate to Y2K?

All these things relate one to another when viewed as an overall picture of world events, whether they took place 4,000 years ago, 2,000 years ago or 200 years ago, or are taking place today.

All events will lead to the climax, the Great Tribulation, which will be ended by the appearance of the Lord in great power and glory.

If we understand this correctly from a Biblical perspective, we can see the puzzle coming together in Bible prophecy as God has given it to us to identify things to come.

Or, to say it in Biblical terms, *"The Revelation of Jesus Christ, which God gave unto him, to shew unto his servants things which must shortly come to pass; and he sent and signified it by his angel unto his servant John"* (Revelation 1:1). ■

CHAPTER FOURTEEN

Biblical Conspiracies

This chapter will show that the prophesied global unity isn't coming about due to the leadership of a brutal dictator, as many believe, or through conspiracy by such organizations as The Rockefeller Foundation, The Council On Foreign Relations, The Bilderbergers, and The Club of Rome; it is springing from economic necessity.

Since the fall of the Iron Curtain, the spirit of democracy is moving to the East and it won't stop at China, which is now practicing the early stages of Roman democracy and capitalism.

The few countries which refuse to fall in line are suffering greatly. Communist Cuba, some say, is economically 40 years behind and the people of communist North Korea are literally starving.

Thus, we see democracy as the answer for a united world, thereby fulfilling Bible prophecy as the nations unite to establish the final Gentile world empire.

Y2K is playing a part by assisting the nations in becoming more and more dependent upon one another.

It is rather amazing that the Bible uses the word "conspiracy" many times.

Israel's history is punctuated by conspiracies throughout the reigns of its various kings. Some of these were positive conspiracies and others were negative.

Joseph

Before Israel became a nation, the sons of Jacob conspired against Joseph. We read in Genesis 37:18, *"And when they saw him afar off, even before he came near unto them, they conspired against him to slay him."*

What a terrible and shocking testimony of Jacob's family! They were from the stock of Israel, from which God would bring forth the twelve tribes destined to rule the world.

Moreover, from the nation of Israel would come forth the Savior, the Messiah of Israel and the Redeemer of the world. Yet this family was anything but a model for such a high calling!

David and Absalom

Another terrible ordeal for Israel began when King Saul conspired to have his son-in-law, David, assassinated. And later, when David became king over all of Israel, his own son, Absalom, conspired against him!

The Prophets

Another conspiracy is recorded in Ezekiel 22, beginning with verse 23, *"And the word of the LORD came unto me, saying,*

"Son of man, say unto her, Thou art the land that is not cleansed, nor rained upon in the day of indignation.

'There is a conspiracy of her prophets in the midst thereof, like a roaring lion ravening the prey; they have devoured souls; they have taken the treasure and precious things; they have made her many widows in the midst thereof.

"Her priests have violated my law, and have profaned mine holy things: they have put no difference between the holy and profane, neither have they shewed difference

*between the unclean and the clean, and have hid their eyes
from my sabbaths, and I am profaned among them.*

*"Her princes in the midst thereof are like wolves raven-
ing the prey, to shed blood, and to destroy souls, to get dis-
honest gain.*

*"And her prophets have daubed them with untempered
morter, seeing vanity, and divining lies unto them, saying,
Thus saith the Lord GOD, when the LORD hath not spoken.*

*"The people of the land have used oppression, and exer-
cised robbery, and have vexed the poor and needy: yea, they
have oppressed the stranger wrongfully"* (verses 23–29).

What a shocking conclusion the Lord God, the Creator of
heaven and Earth, has to make about His own chosen
nation, the Jews! The terrible wrongdoing includes all the
people—the religious establishment, the political establishment and the social establishment.

Particularly important to note is that this conspiracy was
primarily the work of the prophets. These were the very
people who should have been the most helpful, admonish-
ing the Jews to keep the law, to do good, to obey the com-
mandments and to practice righteousness.

But these prophets were apparently the most corrupt.
They had only one thing in mind: their *own* well-being!

Because the prophets were so corrupt, it is not surprising
that the government and the rest of the people were equally
corrupt.

Sin always breeds sin and the end thereof is destruction.
The conspiracy of the prophets brought darkness upon the
nation because it originated with the father of lies, the great
deceiver, who wants all men to be lost for all eternity.

Jehoiada

A positive conspiracy took place around 885 B.C. when the
priest Jehoiada conspired against Athaliah the queen and

made Joash, a true descendant of the house of David, king over Judah. We read about it in 2nd Chronicles 23:1–3, *"And in the seventh year Jehoiada strengthened himself, and took the captains of hundreds, Azariah the son of Jeroham, and Ishmael the son of Jehohanan, and Azariah the son of Obed, and Maaseiah the son of Adaiah, and Elishaphat the son of Zichri, into covenant with him.*

"And they went about in Judah, and gathered the Levites out of all the cities of Judah, and the chief of the fathers of Israel, and they came to Jerusalem.

"And all the congregation made a covenant with the king in the house of God. And he said unto them, Behold, the king's son shall reign, as the LORD hath said of the sons of David."

When reading this entire chapter, it is fascinating to see that Jehoiada and his friends totally ignored the enemy, in this case, Queen Athaliah. They just did what was right. They based their action on the Word of God, *"Behold, the king's son shall reign, as the Lord hath said of the sons of David."*

Then they proceeded to make all the necessary preparations without planning to destroy the wicked queen. They ignored the enemy!

"Then they brought out the king's son, and put upon him the crown, and gave him the testimony, and made him king. And Jehoiada and his sons anointed him, and said, God save the king" (verse 11).

The action of "doing the right thing" caused the self-exposure of the enemy, *"Now when Athaliah heard the noise of the people running and praising the king, she came to the people into the house of the LORD:*

"And she looked, and, behold, the king stood at his pillar at the entering in, and the princes and the trumpets by the king: and all the people of the land rejoiced, and

sounded with trumpets, also the singers with instruments of musick, and such as taught to sing praise. Then Athaliah rent her clothes, and said, Treason, Treason" (verses 12–13).

Against Principalities

This should be a lesson to us regarding our "enemies." We, too, often fight with "flesh and blood" by identifying all kinds of names and organizations who are supposedly enemies of the Gospel and the work of the church. The Apostle Paul warns against this type of activity in the church and writes to the Ephesians, *"For we wrestle not against flesh and blood, but against principalities, against powers, against the rulers of the darkness of this world, against spiritual wickedness in high places"* (Ephesians 6:12).

It's interesting to read the next two verses, which instruct us to do one thing: "stand."

"Wherefore take unto you the whole armour of God, that ye may be able to withstand in the evil day, and having done all, to stand.

"Stand therefore, having your loins girt about with truth, and having on the breastplate of righteousness..."

It is not our task to attack or defeat the enemy, which we have no power to do any way, but to simply believe in Him who has saved us with His own precious blood and trust in Him. That is standing on the solid and sure foundation of the Gospel!   ■

CHAPTER FIFTEEN

The Great Conspiracy

As we've already stated, the Y2K problem has a tremendous potential to prepare the pathway for the global society of which the Scripture speaks so often. A look at three subjects will help us to understand this:

1) Global military conspiracies

2) Global economic conspiracies

3) Global religious conspiracies

Global Military Conspiracies
Practice for the global military conspiracy took place during the First and Second World Wars. These wars served as testing grounds on which the world united against one specific danger.

The people of the world agreed almost unanimously that Hitler's Germany threatened world peace. They had to unite to oppose him and end the destructive power structure.

World War II was fought primarily against Nazism. The great opponent of the National Socialists (Nazis) was communism. With the help of capitalism, communism came out victorious. The results, however, were unexpected: The war was won but peace was lost. For over 45 years, the cold war targeted capitalism against communism, communism against capitalism. The nations of the world divided themselves into one camp or the other.

When the world united against Iraq's Saddam Hussein in 1991, Russia gave its blessings. For the first time, a truly global military force was established under the auspices of the United Nations, led by the United States. Much has been said and written about the Gulf War, but the end results may not have justified the means. Saddam Hussein was forced out of Kuwait but Kuwait's family dictatorship continues to be the same today as before the war and that was most certainly not the object of the war. The U.S.A. is not supposed to support dictatorships.

We have prided ourselves in assisting nations and peoples in gaining independence under democracy with human rights attached. But today, Kuwait still disregards basic human rights and real religious freedom does not exist. Christians, in particular, are still suffering.

The leader causing the conflict was not removed. In 1999, Saddam Hussein still controlled Iraq and threatened the stability of the Middle East and the world.

Progress Of Communism

In its early stages, communism, under the leadership of the Soviet Union, made great progress. The USSR put the first satellite and the first man into orbit. Furthermore, the USSR orbited and assembled the first workable space station called "peace" (Mir). Recently, Russia initiated the successful launch of the first section for the world's first international space station. But communism has one major flaw. By definition, it does not encourage personal gain. The elite of the nation, the intellectuals, the hard-working businessmen, have been expected to operate without the benefit of incentives. Thus, communism is digging its own grave.

The system which promised paradise on Earth has become a system of oppression and has finally ended in poverty. The fall of communism, however, has opened the door to globalism.

Global Economic Conspiracies

The global "conspiracy" we are discussing is not the result of the United Nations, Bilderbergers, CFR, Club Of Rome, and a host of others as stated earlier. Rather, it will come about because of a natural economic development.

If the United States cut off all economic ties to the rest of the world, other countries would naturally cut ties with the U.S. as well. This would cause, almost instantly, a recession which would quickly lead to a depression much worse than the one experienced in 1929.

The United States produces more food than it can use; subsequently, our surplus must be sold to other countries.

If we could not sell that food, a large number of farmers would be forced to file bankruptcy within a year.

Farm machinery would remain unsold, factories would close, and many Americans would lose their jobs.

The people who lost their jobs would not be able to pay their mortgages for their houses or their loans for their cars.

It doesn't take a Ph.D. in economics to know what kind of chain reaction would take place.

International Investment

But there is more. In the economic field, we are actually dependent upon foreign countries, just as, of course, foreign countries are dependent upon us. An article in our local daily newspaper, *The State*, dated April 23, 1998, reported:

> Foreign firms are creating jobs in the United States five
> times faster than American-owned companies, according to a

study by an international trade association.

The Organization for International Investment, a Washington trade association for 60 foreign firms with U.S. operations, found the largest concentration of such jobs in California. Nevada was the leading state in job growth from these investments. Americans working for the subsidiaries of foreign corporations increased from 2.03 million in 1980 to 4.9 million in 1995, the latest year for which statistics are available, the trade association said."

—*The State*, 4/23/98, p. B-9

This should settle our question about international co-operation. In simple terms, we can't exist independently.

The question is often asked, "Should Christians support or oppose the global economy?"

Let me be brief and frank: There is no option. We can't oppose it. The moment we go to a supermarket, buy a car, or deal with a bank, we are supporting globalism.

Many of us remember how President Reagan, during his eight-year administration, tried desperately to promote Americanism with the slogan, "Buy American."

That was easy to say but impossible to do. One smart reporter took the time to do research on a Japanese car and found that over 50% of that car had been manufactured in the United States. He then did the same research on an American car and the result was that over 50% of the car had been made in a foreign country. This clearly illustrates that nationalism is out; globalism is in!

Global Religious Conspiracies

Contrary to the global military and global economic conspiracies, the global religious conspiracies are more true to their title. Anyone who pays attention to the media, particularly Christians, already realizes that our days are numbered. Christianity boldly proclaims there is only one way to heaven; it is through the Lord Jesus Christ. Man cannot reach God unless he comes through the one Mediator, which is the man Jesus Christ. Let's quote 1st Timothy 2:5, *"For there is one God, and one mediator between God and men, the man Christ Jesus."*

We, fundamental Bible-believers, boldly proclaim the words of Jesus. But this philosophy is totally unacceptable to the global world. Hundreds, if not thousands, of religions have their own peculiar administration, practice, and beliefs.

Bible-believing Christians will never deviate from the eternal Word of God, and we continue to proclaim loudly and clearly that only Jesus saves, and Jesus is coming again.

This message, however, contradicts all other religions which proclaim that their faith is the real way. Thus, the world requests us to be tolerant, to compromise, and to think "ecumenically."

While I'm writing these lines, countless people the world over are trying to create a "formula" acceptable to all people of all religions.

No doubt, the most powerful religious leader in the world (who incidentally is also a powerful political leader), the Roman Catholic pope, continues to proclaim that *unity of religions* is the key to world peace.

Global Worship
How and when the global religious conspiracy will take effect is not known, but the Bible makes it clear that *"...all the world wondered after the Beast. And they worshipped the dragon which gave power unto the Beast: and they worshipped the Beast, saying,*

"Who is like unto the Beast? who is able to make war with him?" (Revelation 13:3–4).

This prophecy must yet be fulfilled. Verse 8 confirms, *"...all that dwell upon the Earth shall worship him...."*

Nebuchadnezzar's Image

As we discussed in Chapter 8 of this book, Daniel chapter 3 offers a preview of world religion. The first world ruler, King Nebuchadnezzar, created a golden image with the express purpose that the entire world would worship the image at the precise time when the music played, *"Then an herald cried aloud, To you it is commanded, O people, nations, and languages, That at what time ye hear the sound of the cornet, flute, harp, sackbut, psaltery, dulcimer, and all kinds of music, ye fall down and worship the golden image that Nebuchadnezzar the king hath set up:*

"And whoso falleth not down and worshippeth shall the same hour be cast into the midst of a burning fiery furnace.

"Therefore at that time, when all the people heard the sound of the cornet, flute, harp, sackbut, psaltery, and all kinds of music, all the people, the nations, and the languages, fell down and worshipped the golden image that Nebuchadnezzar the king had set up" (Daniel 3:4–7).

Today, the end-time image is being created and this image differs from all previous ones. Just one verse in Revelation 13 explains the ultimate image and its authority, *"And he had power to give life unto the image of the Beast, that the image of the Beast should both speak, and cause that as many as would not worship the image of the Beast should be killed"* (verse 15).

CHAPTER SIXTEEN

Prophecy In Y2K?

Although the prophetic Word does not specifically mention the Year 2000, we believe that an event such as Y2K, causing a global information campaign of unprecedented proportion; nevertheless, has prophetic significance.

We have already learned that Y2K concerns virtually the entire world. There is no way out of the problem; we all have to participate.

Global Emergencies

Now we are coming to the point I wish to reinforce; namely, the global emergency. In case of an emergency, every hindering element will be taken out of the way. At this point, there is much disagreement between the nations of the world. The U.S. accuses the European Union of protectionism, and the European Union cries "foul" when it discovers that America's major corporations are receiving billions of incentives from the government.

TIME magazine, November 16, 1998, featured a special report titled "Fantasy Islands And Other Perfectly Legal Ways That Big Companies Manage To Avoid Billions In Federal Taxes." Note the following excerpt:

In 1934, the country was in terrible shape. President
Roosevelt launched an agency to create jobs by offering
loans, grants and long-term guarantees to exporters, in hopes
of getting the country out of the darkest years of the

Depression. Those days, of course, are long gone, but the
Export-Import Bank lives on and guarantees billions in loans
to aid huge corporations.

The Boeing Company of Seattle is the largest recipient of
Eximbank guarantees. From 1990 through 1997, Boeing
received $11 billion, most of it in the form of long-term
guarantees to finance aircraft sales to countries world-wide.
During that period, Boeing's single largest customer was
China.

Other companies that benefit: Caterpillar; Bechtel; At&T;
Foster Wheeler; Westinghouse.

Your cost $4.3 Billion.

—Time, 11/16/98, p.83.

On page 84, the article had this to say about Allied Signal:

A company that thrives ... Over the past five years, Allied's
profits nearly tripled to $1.2 billion. For the entire period, the
company earned more than $4 billion. ... but is still on wel-
fare. During that same period of soaring profits, Allied collected more than $150 million in state and federal corporate
welfare...from multiple sources ... Allied gets export subsidies, loans and guarantees overseas, breaks on real estate
taxes, federal research contracts and incentives to build new
offices.

About dam building and its benefit in the beginning of the
1900s, the article stated on page 86:

In 1902, Congress passed the reclamation act to build dams
and irrigation canals to supply water to small farmers and
their families. The intent of both Congress and President
Theodore Roosevelt was to help out farmers cultivating 160
acres or less. Roosevelt's first reclamation chief declared the

law was to help "a man with a family" and was not to aid corporations.

1998 subsidized water now flows to scores of corporate farms in the American West. Water once earmarked for struggling family farmers goes to agribusinesses the size of entire cities. Big farmers buy the water at a fraction of its real cost.

Beneficiaries include:
- A 22,000–acre cotton ranch
- A Japanese drugmaker
- A multimillionaire potato farmer
  Your cost: $5 Billion.

And about General Electric, TIME reports:

One of the world's best-run and most successful companies, GE has moved smartly to become a global powerhouse.

Bringing in the bucks ... Over the past 11 years, GE's profits rose 228%—from $2.5 billion in 1986 to $8.2 billion in 1997 ...

But still on welfare ... GE gets export subsidies, tax credits, loan guarantees, government-research contracts and federally provided insurance for overseas projects ... and still cutting jobs ... In 11 years, GE has cut more than 120,000 jobs, reducing its work force nearly one-half.

These few excerpts show capitalism and socialism working hand in hand rather successfully.

Eurecom, a monthly bulletin of the European Union, reported in its December, 1998 issue:

In its "1998 Report on U.S. Barriers to Trade and Investment," the European Commission finds that European companies continue to face significant obstacles when trying to export to, or invest in, the U.S.

Largely an update of last year's list, this 14th annual report is a 53–page inventory of problem areas covering tariff barriers, non-tariff barriers, investment-related measures, intellectual property rights and services.

The Commission's critique is set against a contrasting background:

On the one hand, the E.U. and the U.S. have undertaken to reduce or, in some cases, to eliminate obstacles that hinder trans-Atlantic commerce.

On the other [hand], current bilateral trade disputes (bananas anyone?) could damage the overall (and indisputably beneficial) E.U.-US trade and investment relationship, as well as the World Trade Organization's (WTO) credibility in trade dispute resolution.

While joint initiatives like the Transatlantic Economic Partnership (TEP) area step in the right direction, especially with barriers of a regulatory nature, they are not enough to prevent new problems from emerging.

For example, an unwelcome development is the introduction of sub-federal selective purchasing laws restricting the ability of E.U. and other companies doing business in specific countries to bid for contracts in various U.S. states and cities.

Such legislation has been adopted by the Commonwealth of Massachusetts in the case of Burma (which was recently struck down by a U.S. Federal Court), as well as by 20 other cities and local authorities.

These measures aim to regulate the behavior of economic operators that are outside U.S. territorial jurisdiction and affect the conduct of normal international relations.

The United States is not the only offender. America has many legitimate reasons to complain about inaccessibility and barriers which the European Union has erected.

We simply publish these excerpts because in the United States, we quite often read about other countries making it difficult for U.S. products to be imported, but we don't usually see the picture which shows that this matter is a two-way street. In the same issue of *Eurecom*, another article reveals another conflict:

> Among the various sticking points detailed in the 1998 U.S. trade barriers report, the E.U. has just referred one to the WTO for resolution: the US' 1916 Anti-Dumping Act.
>
> Following a January 1997 complaint by the European Confederation of Iron and Steel Industry (eurofer), the Commission has investigated the 1916 Act under the E.U.'s Trade Barriers Regulation and concluded that the legislation breaches WTO anti-dumping rules and threatens to disturb trade relations with the US. This is the first time the Trade Barriers Regulation has led to a formal WTO panel request.
>
> In particular, the E.U. believes the 1916 Act violates the WTO Anti-Dumping Agreements since it allows federal courts to impose criminal penalties on importers, exposing them to treble damages. Under WTO rules, the only possible remedy to dumping is antidumping duties alone. Other U.S. infringements of the Agreement include the lack of procedural rules, of a definition (and qualification) of the injury concept and of criteria for the calculation of "normal value."
>
> Further, the E.U. is concerned about U.S. industry's use of the 1916 Act to restrict access to the U.S. market. This has been highlighted in a recent case initiated under the Act by a U.S. steel producer against a subsidiary of a German steel company.

Although the democratic world is aiming toward a free market economy, it is virtually impossible at this time to fully implement it.

So, the arguments between the nations will continue; there will be conferences and negotiations, agreements and proposals, compromises and new laws, but that we may consider as "normal," in the process of globalization.

Unity In War

What happens when an emergency of global proportion arises? To find the answer, let's look at recent history. During the Second World War, the urgency was so great that no nation could afford to heed the letter of the law they had signed after the First World War at the Geneva Convention.

Human rights, national agreements, business contracts, etc., had to take second place.

The first and foremost emergency to be taken care of was to free the world of the acute threat of Nazism. Capitalism was forced to help communism to defeat Nazism. In the case of war, we all know that each person sacrifices his or her own individual identity to patriotism.

Natural Disaster

In case of natural catastrophes, as in war, we see over and over again that a spirit of unity comes about where rich and poor, Catholics and Protestants, Republicans and Democrats, forget their differences to unite in the fight against the prevailing threat of the catastrophe.

As the world unites in the face of Y2K, it is quite natural that new resolutions will be made and old existing ones will be discarded.

The active players will win, the non-active will lose. Y2K will help sort out the "good" from the "bad," the "winners" from the "losers."

When everything is in place, the year 2000 will begin with a more concentrated, more accurate, and a more truly global world. ∎

CHAPTER SEVENTEEN

European Union

As previously stated, Y2K has little, if any, direct relationship to Bible prophecy. But due to the preparation for unity the world over, we may study the Holy Scriptures and use Y2K as a parable or a shadow of the contours of coming developments. Let's look at this in relationship to the coming of the Antichrist.

Rome Rules
Bible scholars generally agree that Rome is the last Gentile world empire. By "Rome," we refer to the entire Roman Empire, whose philosophy and civil laws have ruled the world since the time of Jesus and will continue to do so until He comes again.

From the revived spirit of Rome, a new global empire will be created. There is little doubt in my mind that the center of activities will be the continent of Europe. Slowly but surely, various individual nations are surrendering part of their sovereignty in order to create a solidly unified European Union.

"We Need More Europe"
In the fall of 1998, I visited the European Parliament in Brussels, Belgium as part of a group of students hearing a two-hour lecture on the importance and future of the European Union.

The following information was related to us: The European Union's office is operated by approximately

30,000 employees who support the 626 elected representatives of the European Parliament. All sessions are held in 11 languages, making it 110 possible language combinations, which are being accomplished by 7,500 professional translators.

Besides the Parliament's prime job of establishing laws and regulations for the 15 member states, the European Union deliberately aims to increase the responsibility for the European Union at the cost of national sovereignty over member nations.

To a question filed from the audience relating to the future of independent nations, the answer was given: "We need more Europe and less nationalism."

Membership in the European Union is permanent and irrevocable. No provision is made in the European Union's Constitution for cancellation of membership.

Why Fifteen Stars?

Another question from the audience was, "Why does the European Union flag consist of only 12 stars in a circle and not 15?" *Answer:* The twelve stars was a symbol of unity and harmony during the Middle Ages.

The Union has determined to keep the 12 stars regardless of how many more states will be added in the future. The history of the flag originates with a statue of the "Virgin Mary" seen in Strasbourg, France. She wears a crown composed of 12 golden stars.

Most Bible believers do not need to be reminded that 12 is the number of the tribes of Israel and that there are 12 apostles of the Lamb. Biblical Israel becomes visible in the flag of the European Union.

How can Europe be united? The answer is rather simple. As new laws are being created, old ones need to be eliminated. New laws will unify all member nations.

Biblical Scenario

Daniel wrote down the prophecies about the Gentile world during the endtimes. The Antichrist, described as, *"A king of fierce countenance,"* has *"understanding of dark sentences."* That clearly indicates knowledge of the occult.

Daniel 8:24 then continues, *"He shall destroy wonderfully."*

This seems to be a contradiction. What's so wonderful about destruction? No doubt, this refers to his ability to do away with (destroy) the old established laws and replace them with new laws. Verse 25 continues, *"...by peace shall destroy many."*

Again, a seeming contradiction; peace does not destroy. However, by friendly dialogue, he will democratically eliminate opposition.

One Must Rule

Europe is growing stronger day by day and prosperity will increase in spite of temporary setbacks here and there.

When we compare Europe with the United States, one difference is striking. In the U.S., one man, the president, rules. In Europe, 15 do.

But that is only the beginning. Many more European states will be accepted into the Union as time goes on. Some estimate that before the year 2010, the European Union will be 800,000,000 strong!

The European continent currently consists of 43 nations. When many more nations are added, their leaders will participate in the decision process and, at that time, the desire for one person to be the ruler of all will become quite evident.

How each step of the process will take place cannot be determined at this time. After all, who would have thought in the 80s that the Soviet Union would collapse? No one ever dreamed that world communism would have to make room for democracy.

For that reason, it is not wise to put down a plan regarding the development of the nation because things develop, quite often, differently than we expect.

One thing is absolutely sure: There will be an Antichrist. The Treaty of Rome brought about the beginning stages of the European Union and the Bible makes it clear that the same political system and people who ruled the world when Jesus as born will be in power again in the end stages of the endtimes.   ■

CHAPTER EIGHTEEN

The Computer and the Beast

Realizing a computer's seemingly endless and instantaneous capacity for memory, research, math and statistics, we are beginning to realize that this relatively new thing called the computer is more than we may want to admit.

The computer can potentially become all-knowing. Do we hear the footsteps of the great imitator, the father of lies? Only God, the Creator of heaven and Earth, is omniscient, but now, even in its infant stages, the computer begins to rise to the challenge.

Revelation 13 describes the success of the Antichrist and the false prophet. In verses 14–15, we read,

"And deceiveth them that dwell on the Earth by the means of those miracles which he had power to do in the sight of the Beast; saying to them that dwell on the earth, that they should make an image to the Beast, which had the wound by a sword, and did live.

"And he had power to give life unto the image of the Beast, that the image of the Beast should both speak, and cause that as many as would not worship the image of the Beast should be killed."

The false prophet, who is identified as *"another beast,"* possesses power and authority. Contrary to the first beast, the Antichrist, he does not have the power and authority, but receives it directly from the Devil, the great dragon, the old serpent, also called Satan.

Let's list a few examples:

" ... *the dragon gave him his power, and his seat, and great authority"* (verse 2).

Verse 4, " ... *the dragon ... gave power unto the Beast."*

Verse 5, twice we read the word *"given."* " ... *there was given unto him a mouth speaking great things and blasphemies; and power was given unto him to continue forty and two months."*

Furthermore, in verse 7, *"And it was given unto him to make war with the saints, and to overcome them: and power was given him over all kindreds, and tongues, and nations."*

It looks like the Antichrist is a poor chap; he has nothing because everything is given to him!

This item is interesting in view of the fact that Hitler, the man who plunged the globe into World War Two, was also an insignificant person before he received power. He was just an unknown poor painter with little education and no accomplishment of any sort to claim as his own.

As an Austrian national, a foreigner in Germany, this one of many forerunners of Antichrist successfully deceived the nation and became the most feared man on Earth.

The Little Horn

Daniel chapter 8 offers this description of the Antichrist, *"And out of one of them came forth a little horn, which waxed exceeding great, toward the south, and toward the east, and toward the pleasant land.*

"And it waxed great, even to the host of heaven; and it cast down some of the host and of the stars to the ground, and stamped upon them.

"Yea, he magnified himself even to the prince of the host, and by him the daily sacrifice was taken away, and the lace of his sanctuary was cast down.

"And an host was given him against the daily sacrifice by reason of transgression, and it cast down the truth to the ground; and it practiced, and prospered" (verses 9–12).

This is the Antichrist, the one to whom "was given" everything he has. He is the one who prospers and finally will become the leader of the entire world. In verse 24, we again read this sentence, *" ... but not by his own power ...."*

The False Prophet

The other beast, who promotes the Antichrist, is quite different. He possesses power, as we can clearly see in verse 11, *" ... he had two horns like a lamb ...."*

Verse 12, *"And he exerciseth all the power of the first beast ...."* He *" ... causeth the Earth and them which dwell therein to worship the first beast ...."*

Verse 13, *" ... he doeth great wonders ...."* *" ... he maketh fire come down from heaven ...."*

Verse 14, *"And deceiveth them that dwell on the Earth... he had power to do in the sight of the Beast ...."*

It is important to understand the distinct difference between the first beast, who rises out of *"the sea"* indicating that he originates from among the masses of the people on Earth, and the second beast, who comes *"out of the Earth"* signifying his geographical existence on Earth.

This second beast, the false prophet, promotes the Antichrist and gives orders to them *"that dwell on the Earth that they (the people of the Earth) should make an image to the Beast ...."*

Images In the Bible

Images are reported for us throughout the Scripture. One of the most infamous image is the golden calf, which Israel made, and with which it committed idolatry against the God of Israel. Then there were the images King Solomon bowed down to, which he had tolerated with the wives he married. Israel, indeed, had her share of images throughout history.

For the Gentiles, the most famous, or better said, infamous image, is no doubt the one erected by Nebuchadnezzar, the first Gentile world ruler.

But all of these images, whether in Israel or in pagan lands, they all fit the category the Apostle Paul describes in 1st Corinthians 12:2, *"Ye know that ye were Gentiles, carried away unto these dumb idols, even as ye were led."*

The Old Testament describes an idol maker and his idol with these words, *"What profiteth the graven image that the maker thereof hath graven it; the molten image, and a teacher of lies, that the maker of his work trusteth therein, to make dumb idols?*

"Woe unto him that saith to the wood, Awake; to the dumb stone, Arise, it shall teach! Behold, it is laid over with gold and silver, and there is no breath at all in the midst of it" (Habakkuk 2:18–19).

These idols, by whatever name, were unable to speak, hear or respond. They were dead, unable to do or initiate any functions mimicking man, who is a living soul.

The Global Image

The image mentioned in Revelation 13, however, is different. The false prophet gives the order for man on Earth to *"make an image to the Beast."*

Is this image like unto all other images we read about in our Bible? No, because the next verse states, *"And he had power to give life unto the image of the Beast, that the image of the Beast should both speak, and cause that as many as would not worship the image of the Beast should be killed"* (Revelation 13:15).

This is absolutely amazing! The false prophet has *"power to give life."* Do we see the imitation of God the Creator here? *"And the Lord God formed man of the dust of the ground, and breathed into his nostrils the breath of life; and man became a living soul"* (Genesis 2:7).

The false prophet is the spokesman for the Antichrist.

He is the "word," the imitation of John 1:1 and 3, *"In the beginning was the Word, and the Word was with God, and the Word was God.*

"All things were made by him; and without him was not any thing made that was made."

Until this point, only God had authority to give life and to take life. Here we see the great imitator at work who gives "life" to a manufactured image. Through this created image, the false prophet has power to take life, *" ... and cause that as many as would not worship the image of the Beast should be killed."*

I think by now we all understand which direction we are headed in this discussion; namely, the computer's role in fulfilling the image of the Beast.

This, of course, was unthinkable only a hundred years ago. No such thing as a computer existed and no one had dreamed that the computer would be invented. No one really imagined that we would experience the phenomenal growth of the computer industry.

Our First "Computer"
I recall an incident from the beginning of our ministry, sometime in the early 70s. The ministry started growing from the little basement office in our home to a small addition at the back of the house.

One day the time had come to purchase a calculator.

I scanned the yellow pages, made several phone calls, and got in touch with a salesman at a business machine distributor in Cincinnati, Ohio. He informed me that mechanical calculators were complex, expensive and on the way out.

I listened carefully as he explained the amazing capability of the new calculator he had for sale.

The price was the only thing which caused me to hesitate: It cost almost $200. During those days, that was a lot of money!

The salesman then offered to come by for a no-obligation demonstration of this marvelous electronic calculator. He came and demonstrated the machine's basic four functions. After some negotiation, Midnight Call Ministry was in the possession of this cutting-edge gizmo at the substantially reduced cost of $145!

Faster, Bigger and Cheaper

How fast the electronic industry has developed since then is unbelievable. A little computer watch you can purchase at a retail store for about $25–$30 has about 100 times as much memory and about 20 times the functions that $145 electronic calculator we purchased in the 1970s had.

Even the best computers available today will be surpassed shortly. Each year, sometimes each month, better, faster and more capable computers are being offered to the public and at a lower cost.

Magazines such as *Popular Science* give us a glimpse of new inventions and the latest developments on a monthly basis.

Who's In Charge?

Someone may now object and point out that the image has to be initiated by the false prophet and only then will the image of the Beast be produced. That certainly rings true but we are overlooking the fact that building such an image can only take place if the technology exists.

Is it possible today to build an image that can speak? That is old news. But what has not yet transpired is that authority is not yet in the hands of a computer. Some people claim that the computer is already in charge: I do not agree. No doubt, however, it is in the works.

An example will help us better comprehend this. Let's say you have purchased a ticket to fly from New York to Paris. You arrive at the airport and give your ticket to the check-in counter person. What will he do with your ticket? First he will contact the computer to confirm that the ticket corresponds to its stored data.

Should the computer's information indicate that the particular flight is sold out or that your name is not registered, you most likely won't get on that plane. Although you may have the ticket, in such cases, "The computer is always right."

At this point in time, we are still battling between hard (written) and soft (electronic) copies, but "soft copies" are already winning. When we analyze such developments, we clearly see where they are leading.

A Man and His Word

In "the olden" days, a man was said to be as good as his word. For example, when the Roman centurion asked Paul, "Are you a Roman citizen?" Paul's simple reply of "Yes" was sufficient.

But today, the man's words have been replaced with documents. Certain pieces of paper, tickets or passports are supposed to confirm our words. We can't go to another country or enter the United States without a passport. No matter how honest or trustworthy we may be, immigration

officials will not let us cross borders unless we have a document which confirms that we are who we say we are.

Mind Reading?

Not only will the manufactured image be brought to life by
the false prophet and not only will it be able to speak, but the
image will be able to recognize who is worshipping and
who is not.

There is more. The image will become the ultimate
authority over a person's life. I don't believe it hurts to be
repetitious in this case because of the importance regarding
the authority of the manufactured image. " ... *the image of
the Beast should both speak and cause that as many would
not worship the image of the Beast should be killed."* That
is total religious control!

Churches, mosques, synagogues and other places of so-
called "worship" will be filled to overflowing, but the object
of worship will not be the Lord who became man, died on
Calvary's cross, poured out his life in His blood, was buried,
arose the third day and liveth forever and ever, but the object
will be the image of the Beast who apparently was dead, too,
and rose to life.

This is reported for us in verse 3 of chapter 13, *"and I
saw one of his heads as it were wounded to death and his
deadly womb was healed ...."*

That is another seemingly perfect imitation of the Lord
who gave His life a ransom for many, but he, the Beast,
receives life in order to deceive the entire world into eternal
destruction.

Control Of Commerce

A united world religion is not sufficient to totally control
planet Earth. Just as we now have a global democracy, we
must have a global economic system. The entire industrial

economic might have planet Earth will come under his control, *"And he causeth all, both small and great, rich and poor, free and bond, to receive a mark in their right hand, or in their foreheads:*

"And that no man might buy or sell, save he that had the mark, or the name of the Beast, or the number of his name" (Revelation 13:16–17).

Is this something that is real or is John using a figurative symbolical language? I have no doubt in my mind that this is just as real as Satan giving power to the Antichrist and the false prophet exercising his power to give life to the image of the Beast and, thereby, controlling the one-world religion, in the process, eliminating those who oppose him.

Worship is something that is very personal and to be killed is just as personal. There is no figurative or symbolic language here. Verses 16–17 very clearly speak about real people during a real time. The world has always had, and always will have, the small and the great, the rich and the poor, the free and the bond. There is no symbolic language here. It means all people on planet Earth. ■

CHAPTER NINETEEN

Survivalist

If those who do not worship the image of the Beast somehow escape into a remote area, they still will find it extremely difficult to exist without buying and selling for an extended time.

Many new cults are arising in these last days whose religion is summed up in one word—survival.

Survivalists assume that if they store enough food and ammunition, they will survive. That, of course, is wishful thinking, because any type of survival is only temporary at best. Modern man has become interdependent and cannot exist in isolation.

The survivalist philosophy is basically flawed. In the long run, people will always need medical attention. Those who live in seclusion will at some point have to go to town to get, if nothing else, some salt. It is naive to think that someone can live with his family and a group of like-minded people in isolation from modern society for any extended period of time.

Financial Control

Several years ago, we were made aware of our dependency on modern conveniences such as credit-card accounts. Our merchant account with Visa/Mastercard was audited.

Without taking particular notice of the small print, we accepted credit-card charges for large amounts, such as conferences and tours to the Holy Land. The auditors made us aware that this was not the type of service we originally applied for. We had to make quick adjustments in order not to lose our merchant account. If we had refused to comply with their request, they would have simply closed our account.

That brought us to a rude awakening; the entire ministry was at stake. Approximately 15–20% of our income needed to operate would cease immediately.

Midnight Call and virtually every other organization and business has become dependent on receiving payments through credit-card holders.

Furthermore, our regular bank account is actually a lifeline for our ministry. If our bank were to close our account and, for some unknown reason, we were unable to immediately open a new account with another bank, this ministry would not last seven days. This doesn't just apply to us, but for any other operations, organizations and businesses as well.

Toward Total Control

Now, the false prophet, in conjunction with the living image of the Beast, begins to implement total commercial control of each individual person on Earth.

How does all this relate to Y2K? The year 2000 is a stepping stone on the road to global electronic unity.

The immense volume of computer scientists the world over, each with his or her own ideas, programs and targets, is now being forced to work in unison. That is the good and bad news about Y2K. In summary, the "good" will lead to the "bad." ∎

CHAPTER TWENTY

Faith Versus Facts

W hen analyzing the Y2K problem, we must also take into consideration the extent faith plays in this entire matter. During conversations with Y2K-concerned people, I have noticed that faith plays a major role in their concern. Those who believe that Y2K will result in chaos can instantly quote a volume of infor-mation "proving" that the world will experience a catastro-phe come January 1, 2000.

In one instance, someone gave me an article about Y2K which showed the potential problem facing our nation and the world. While I was reading the report, I made sure to highlight some of the positive aspects which I thought clearly showed that Y2K is relatively "under control."

After showing my underlines to the person, he quickly "destroyed" any evidence I thought I had discovered in the article by offering his additional arguments. We actually continued our debate for several days but we always came up with a diametrically opposed result.

Y2K Negative Versus Y2K Positive

I have concluded that there are two types of people who, while viewing the same matter, are coming to two different conclusions. I have coined them "Y2K negative" and "Y2K positive." In speaking of faith, we are not speaking of the faith that saves, faith in the Lord Jesus Christ, but we are speaking of faith which is expressed through our emotion, or, to name it Biblically, through our soul. Our soul is that undefinable part of our personality which registers, realizes, and digests everything that is visible, tangible and that can be expressed with our emotion. Our spirit, however, born-again, ignores everything that is perceptible to the soul, but rather it connects us directly to our heavenly Father. Our Spirit sees in faith the spiritual things that are heavenly and, therefore, eternal.

The Sword Of the Word

No wonder the Word of God is called a powerful sharp sword which separates the soul and the spirit.

"For the Word of God is quick, and powerful, and sharper than any two-edged sword, piercing even to the dividing asunder of soul and spirit, and of the joints and marrow, and is a discerner of the thoughts and intents of the heart" (Hebrews 4:12).

I repeat, when we are speaking of faith here, we are concerning ourselves with the faith based on our soul. Our faith based on the visible is the faith of the soul, but our faith based on the Word of God, trusting His spirit, is the faith that saves.

Let's examine a political example:

If I am a Republican, then I will generally believe and "have faith in" an article that speaks negatively of Democrats.

This, of course, is true in the reverse as well. A person who accept the Democratic Party's philosophy will eagerly confirm a good report about the Democrats but quickly point out the negative aspects of the opposing party.

Therefore, if you believe that the Year 2000 will usher in unprecedented catastrophe, you will find, see and believe virtually anything that confirms your belief toward Y2K negative.

Will Science Win?

Although I lean toward the positive category, I don't consider myself unconditionally Y2K positive.

Here are some reasons why: Those who deny the potential danger of Y2K simplify the matter by summarizing that technology was invented by man and man always was and always will be the master of his creation. This attitude presupposes the solving of problems because it has been done in the past.

Some quote statements made during the beginning of this century when the industrial revolution took hold of the Western civilization, primarily under the leadership of the United States. The speed of cars, they said, is limited to about 35 m.p.h. Higher speed would destroy the very composition of man's body.

When some of the leading scientists of the world were working in the United States to create the nuclear bomb, speculations and rumors were rampant. Many claimed that once a nuclear reaction was started, it wouldn't stop until planet Earth was destroyed.

Many other predictions relating to modern technology have been proven wrong. Man, indeed, remains the master of his creation.

Irreversible Y2K

However, Y2K presents mankind with two irreversible facts:

1) *Y2K is unprecedented.* We have never had computers before and, unfortunately, we do not know for sure how an interdependent computer system will act in the event some components within the system do not recognize "00" to be the year 2000. This is new; this is unique; this is a first.

2) *Y2K is unforgiving.* The date January 1, 2000, is set in concrete. We can't get an extension; we can't postpone it. This fact, again, reinforces the uniqueness of the Y2K problem. We can't compare it with the building of the pyramids or the manufacturing of an A-bomb. Schedules had to be met, time was scarce, and uncertainties abounded, but each and every project in the past has had the option for an extension. Y2K does not.

Summary

Based on those facts, I cannot unconditionally align myself with either camp. I find it completely unnecessary to adapt the survivalist attitudes which call for stockpiling money, food, and fuel. However, neither can I accept the "couldn't care less" attitude of simply waiting to see what will happen.

Our ministry has made the necessary preparations. We have tested the various operation systems and programs.

And, according to System Manager Joel Froese, Midnight Call Ministry is Y2K compliant, although minor inconveniences may occur.

Later in this book, Joel will answer many questions regarding the technology of computers, computer chips, their action and reaction, to make it easier for the average layman to understand.   ∎

CHAPTER TWENTY-ONE

How Ready Is U.S. Society?

L et us now investigate three important issues:

1) What has been done about Y2K?
2) What can be done about it?
3) What must be done?

What Has Been Done About Y2K?

Actually, very much has been done and accomplished. It is unthinkable that at the time when this book goes to press that there are businesses which do not know about Y2K.

Y2K has been discussed among experts since the early 80s, when computer scientists and programmers saw it coming. They knew this news must be made public knowledge.

U.S. Government

The Office of Government Policy has released the now well-known White Paper which summarizes the Federal Government's action to prepare computer systems to operate in the year 2000. This came to my desk dated August 1998:

This White Paper discusses the major developments,

activities, progress, and the status of Federal agencies'
efforts in updating their computer systems and networks for
continued operation when using "Year 2000" or "Y2K"
dates. It follows an earlier General Services Administration
(GSA) Y2K White Paper that focused on the Federal
Acquisition Regulation's (FAR) procurement direction for
acquiring Year 2000 compliant products and services.

While the Y2K challenge remains daunting, there are
areas of positive action and encouraging results.

Recent Congressional action, including the establishment
of the Senate Special Committee on the Year 2000
Technology Problems, reflects Congress' growing concern
and involvement in the Y2K issue. One very encouraging
note is that the Senate Special Committee has indicated a
significant shift in the long-standing "policy"—that Y2K-
correction funding must come from existing agency appropriations. Chairman Senator Bennett, in announcing the
Committee's agenda, estimated that the costs to fix the government's systems will exceed $10 billion—more than twice
Office of Management and Budget's (OMB) latest estimate—and indicated that "significant appropriations" of
Y2K funding are likely. Already, over $86 million in supple-
mental appropriations has been authorized to pay for Year
2000 fixes. At this writing the Senate Appropriations
Committee has also created a $2.25 billion emergency
reserve fund that agencies will be able to access. OMB
believes that available and planned budgets will provide sufficient funding to address Y2K corrections.

The Administration has taken steps to oversee agencies'
activities and help agencies to ensure timely Y2K fixes. The
President has designated Y2K corrections for tracking as a
major government wide management performance measure.
Further, on February 4, 1998, the President issued Executive
Order 13073, "Year 2000 Conversion" which, among other

things, established the President's Council on Year 2000 with the Chairperson directly reporting to the President. The Chairperson, John Koskinen, sees Y2K as managing a three-tier problem: gaining Y2K control over the agencies' own systems; making sure those systems work with external data-exchange partners' systems; and managing and minimizing the global/international impact of Y2K problems.

In addition the Office of Personal Management (OPM) is providing two new tools to attract and maintain Y2K human resources. First, if requested, OPM will waive the restriction on dual compensation reductions making it easier to re-employ retirees with required programming skills or system knowledge. OPM has also authorized agencies to offer premium pay for employees working to correct Y2K programs.

GSA also increased its leadership role by acting as facilitator and collection point for government-wide Y2K guidance and best practices. GSA's Office of Government-wide Policy, at the request of the Chief Information Officer (CIO) Council Subcommittee on the Year 2000, developed and maintains a comprehensive repository of Y2K information and developed a Y2K COTS product compliant database. The web site, known as the Year 2000 Information Directory, acts as a clearinghouse for information on the Y2K computer challenge and is located at http://www.itpolicy.gsa.gov. This represents a wealth of information and guidance on the status of the Federal government's actions and plan of attack to solve the Year 2000 problem. Individuals and agencies seeking information on the Y2K issue are strongly encouraged to visit this site.

Representing the largest single information technology problem ever faced, Y2K demands our undivided attention and focus. Agencies are taking aggressive action to identify and correct potentially disruptive and damaging Y2K systems programming problems. As noted in OMB's most

recent (May 15, 1998) Quarterly Report, Progress on Year
2000 Conversion:

"Overall, the Federal government continues to make
progress in addressing the Y2K problem—but the rate for
some agencies is still not fast enough. OMB categorizes
agencies into one of three tiers based on evidence of progress
in their reports.

Tier 1 comprises agencies where there is insufficient evidence of adequate progress.

For agencies in Tier 2, OMB sees evidence of progress,
but also has concerns. The remaining nine agencies in Tier 3
are making satisfactory progress.

Although 71% of the systems in Tier 3 are compliant,
only 31% of the systems of the Tier 1 agencies are compliant. It is critical that those agencies most at risk devote more
management attention to the problem in order to ensure that
solving the year 2000 problem is the agency's highest information technology priority."

Much has been done since the White Paper was released,
and almost daily, system managers, branches of operation
etc., report being Y2K compliant.

Our own investigation reveals that virtually all government and private sectors of the industry are currently Y2K
compliant or expect to be so before the Summer of 1999.

Boeing Industry
The United Kingdom's publication *Flight International*,
September 15, 1998, reported the following about Boeing:

The Year 2000 software nightmare scenario is like something from the 1951 science fiction classic The Day The
Earth Stood Still in which an omnipotent alien paralyses the
world by shutting down every electrically operated device

for an hour.

If Boeing is right, nothing like this scale of disruption will occur as the world's aircraft computer-based systems move into the next century. The worst that can happen, claims Boeing, will be delayed dispatch or—the case of an aircraft with an early inertial navigation system (INS)—pre-flight procedures may not be able to be carried out.

"We started in 1993 to address the Y2K issue for airborne systems," says Boeing Commercial Airplane Group (BCAG) aircraft systems

Vice President Tim Fehr. All Boeing airborne systems were evaluated to identify any possible defects or adverse impacts from "date-rollover" processing at midnight on 31 December, 1999.

The company's Airborne Systems Year 2000 Team study included all Boeing airliner models now under development, in production, or in use, including military derivatives. It also included Boeing-approved tools required for normal customer airlines' operation.

Part of Boeing's problem was knowing where to stop. Legally, it is obligatory that, for aircraft still under warranty, all Boeing-designed or created software will function regardless of the year.

—*Flight International*, September 15, 1998, p. 60

Airbus

Boeing's European competition Airbus released the following information:

Airbus Industry set up a task force several years ago to examine the potential impact of the Y2K software problem on its aircraft. The consortium has concluded that there are only a few minor issues that will affect its products. It does, however, warn that it cannot take responsibility for items over which it does not have control.

The Y2K issue at Airbus is being tackled by a 14-strong team headed by project manager Terry Whiting, based at the consortium's Toulouse headquarters. The team is dealing directly with the partners—Aerospatiale, British Aerospace, Deutsche Aerospace Airbus and CASA—as well as customers, vendors, legal issues and the media. Airbus is aiming to have secured Y2K compliance by 1 January, 1999, for every system, service and electronic device linked to its products or its business in general.

John Walker, deputy Vice President of general engineering at Airbus, has been tasked with the product side of the investigation. He says that after at least two years of work the team is now confident of the status of its findings. "The few areas where there are issues are either not business critical or there is a well-defined backup system available," says Walker, adding that software checks have revealed no areas of significant concern

—*Flight International*, September 15, 1998, p. 62–63

Federal Aviation Administration
Recent reports coming to our office from the Y2K negative camp warn consumers not to book flights between December 31, 1999 and January 1, 2000. Some reports actually claim that all airlines have cancelled their flights during that time. Here is the answer coming from the authority, the Federal Aviation Administration.

FAA (Federal Aviation Administration) Ready
Despite intense domestic criticism of its lateness in getting to grips with the year 2000 computer problem, the U.S. Federal Aviation Administration sees itself leading the world's aviation industry safely into the new century.

First, the FAA has to get its own house in order.

Administrator Jane Garvey has acknowledged that the agency was seriously late in beginning its Y2K programme, but the FAA now believes it is ahead of other U.S. government departments in tackling the problem.

As recently as February, the FAA was seven months behind schedule; now it claims to be "on track" to complete renovation of its systems by the 30 September deadline.

"We are looking at having 90% of all systems certificated by 31 March, and the remaining 10% by 30 June," says the FAA. The agency's updates on its progress continue to meet with skepticism, however, not least because of its poor record of bringing large software-intensive projects in on time.

Y2K War Room

While individual business lines are responsible for repairing their systems, the programme office operates a "war room" providing independent verification and validation. Long explains why: "How do I know that what the people in the field are telling me is accurate, and that their systems have been repaired according to the correct process? I don't trust anybody at this point, hence the need for independent verification and validation."

Although the task facing the FAA has been formidable, Long says it has been more of a management, than a technical, challenge. Some 209 mission critical systems, plus a larger number of less critical computers, have had to be assessed and either renovated or replaced.

The FAA struck lucky when tests of the elderly, but vital, host computer used at its en route control centers showed that the 1975–vintage machine does not have a Y2K problem, but instead will be unable to cope with the date rollover to 1 January, 2007.

—*Flight International*, 9/15/98, p.66–67

Air Force

Combat aircraft are unlikely to catch the Millennium bug.

Having produced more digital-era fighters than any other manufacturer, Lockheed Martin should know better than most of its rivals whether the world's combat aircraft will be grounded at the dawn of the new century.

"We are not aware of any airborne issue that has to be resolved," says Robert Elrod, Vice President for the F-16 programme.

The U.S. manufacturer has delivered over 4,000 F-16s so far, making it the most widely used fighter to feature digital computers.

"We've been through the aircraft pretty thoroughly," says Elrod. Lockheed Martin's review was limited to contractor furnished equipment, for which the company is responsible. Ensuring government furnished equipment is Y2K capable is the responsibility of the U.S. Air Force's F-16 system pro-gramme office (SPO). "We've discussed it with the SPO, and they are not aware of any unresolved issues," he says.

A review of ground-based support equipment such as training devices, automatic test equipment (ATE) and logis-tic system software revealed a "reasonably small subset of issues," mainly with the depot-level ATE. Solutions, now under way, are "...relatively straightforward and small in number," says Elrod.

"Most airborne systems are not calendar based, and are most likely to use elapsed time," he explains.

The real test, Elrod admits, will be when the date rolls over.

"We will work with customers to deal with any problems as they occur," he maintains. Meanwhile, the company has made immunity from date rollover problems a specification requirement for any aircraft delivered from 2000 onward.

Flight International, 9/15/98, p.68–69

U.S. Small Business Administration

The U.S. Small Business Administration asked this question in a half-page ad in *The New York Times*, "Are You Y2K OK?" Part of the ad reads:

The Year 2000, or Y2K, date change is a danger to every business. The President's Council on Year 2000 Conversion and the organizations listed here are committed to helping small and medium-sized companies prepare for January 1, 2000.

Business and public awareness has been progressing rapidly. Anyone unsure about Y2K is invited to call a toll-free number for further help or information.

The American Association of Retired Persons featured an article, "Taming the Millennium Bug" in its October 1998 magazine:

The computer bug that threatens to put the bite on systems everywhere on January 1, 2000, may not harm us as much as some people think.

But then again it might.

Such is the conflicting welter of expert opinion that, until the clock strikes the appointed hour, no one will know for sure whether the bug will just be an annoying gnat or the carrier of a dreaded disease.

Older Americans can take comfort, however, in the fact that work under way now should assure uninterrupted delivery of key services. And everyone should be aware that they can increase their protection by taking some simple precautions as the time approaches.

"Benefit payments will continue uninterrupted in the new century." says Kathleen Adams, who heads Social Security's massive computer fix, states flatly.

Social Security

The Social Security Administration got a Y2K wake-up call in 1989 when its computers, doing calculations involving dates beyond 2000, began to misfire. Partly as a result, of its early start at repairs, SSA gets a grade of "A" for its Y2K initiative from the House Government Management, Information and Technology Subcommittee that is tracking Y2K progress.

"We have 308 mission-critical systems, and over 95% of them are remedied," says the SSA's Adams.

But it's not enough for SSA to be ready. The 50 million Social Security payments made each month are actually disbursed by a little-known Treasury office called the Financial Management Service (FMS).

—page 3, 20, 22.

AARP, however, warns about Y2K scams:

Scams May Be Worst Y2K Snag

No Y2K scams have been reported yet, but consumer experts warn that con artists may try to take advantage of people's fears.

"Absolutely, there is a potential for fraud," says Adrienne Oleck, a consumer protection attorney at AARP. "And where there is a potential, I would assume there will be people looking to take advantage of the opportunity."

Her advice: Be wary of unsolicited offers of Y2K assistance. If you have concerns about a service or product, take those concerns directly to the provider

—p.20

Pessimist Verses Optimist

The *Futurist* magazine, September 1998, confronts two diverse opinions:

As frequently happens when a disaster is predicted, the Y2K problem has both its Pollyannas and its Cassandras.

Economist Edward Yardeni of Deutsche Morgan Grenfell, a self-proclaimed "Y2K alarmist," has increased his estimate of the probability for a recession from a 40% likelihood to 60%. The economic fallout could be as disruptive as the recession of 1973–1974 resulting from the oil crisis. At that time, U.S. gross domestic product dropped 3.7%.

"We should prepare for a similar fall in 2000," Yardeni wrote in *The Wall Street Journal* (May 4, 1998).

"Furthermore, a 2000 recession is bound to be deflationary. The U.S. may experience a $1–trillion drop in nominal GDP and a $1–trillion loss in stock market capitalization."

Yardeni suggests that financial markets take a "Y2K holiday" the first week of January 2000 while infotech workers get their systems up to speed.

Meanwhile, governments should contribute to a Y2K emergency fund of at least $100 billion to spend on last-ditch repair and replacement of key computer systems around the world.

But will the fallout be so disastrous? Writing in the same edition of The Wall Street Journal, David Wessel recounted other potential financial disasters that the world has successfully averted.

"Two years ago, the budget tussle between President Clinton and Congress shut down the government for 22 days; no recession ensued," Wessel notes.

"In early 1996, a storm paralyzed large parts of the country for a week; the economy rebounded when the roads were cleared.

"The United Parcel Service strike disrupted shipments last summer; the lasting effects were very small."

The Y2K problem has already been recognized, and though the remedies are not easy, people know what to do and, for the most part, are doing it before the year 2000 arrives.

The situation is not like the Titanic sinking, Wessel maintains: "If the captain of the Titanic had seen the iceberg approaching, he would have steered away from it."

Sources: *Y2K—An Alarmist View* by Edward Yardeni
"Year 2000 Is Cost, But Not Catastrophic" by David Wessel, *The Wall Street Journal* (May 4, 1998, p.18).

Technology Coming Of Age

Popular Science magazine, October 1998, heads an article relating to Y2K, "When the Dust Settles, Year 2000 Will Be Remembered As Technology's Coming of Age:"

A sidestep approach, being employed on a lesser scale, is a 28–year time bridge. The 1972 calendar mirrors 2000's exactly. By setting computers back 28 years, programmers can ensure they operate in the same way.

Often called encapsulation, this method works as long as the actual year doesn't influence the functionality of the program, such as with a key-card access system that uses only day-of-week input.

If the system interacts with external sources, programmers write a software bridge that subtracts 28 years from incoming data and adds 28 years to outgoing data.

According to Hoffman, a time bridge will likely be used to fix Autodin, the messaging system used by the military to send classified e-mail to and from the battlefield.

Years mirror each other in this 28–year window until 2100, so the problem doesn't go away forever with this approach either.

Though windowing and encapsulation are the most common alternative approaches, there are several others that squeeze century information into the existing six-digit space. One method involves making the first digit a century field (where the number 1 equals 19, 2 equals 20, and so on), the next two a year field, and the last three a Julian day-of-the-year field. The benefit of this method is that file size is not affected. The downside: Programmers still have to find all the date fields.

For vendors like the Software Factory, the IT community's reliance on alternative approaches is a boon. "The problem is not going away on January 1, 2000," says Jerry Hill, co-owner of the Software Factory, whose employees all went through a month-long unpaid "boot camp" to learn Cobol before being hired. "We'll have a strong Year 2000 business into the next century."

But programmers on the front lines of the Year 2000 fix are not looking that far ahead just yet—there's still a lot of work to be done before January 1, 2000.

According to Hoffman, the Army's systems are now more than 80% compliant—and on schedule to be completely fixed by the end of this year—but the organization will spend most of next year testing them. "Fixing the code is a small facet," he says. "Testing and implementation take longer than actual code renovation."

That has Martin concerned. "Even though organizations will likely address all their critical systems before the century change," he says, "they may not have time to do proper testing. So the question becomes, Do we really know how our computers run?"

Bob Reinke is convinced we don't. "Year 2000 is going to set the industry back 30 years," he says. "People are going to blame everything on computers for a long, long time." Hoffman is more optimistic, but does worry that word is not getting out to small businesses.

"Walk into the local grocery story," he says, "and ask

them if they've checked their inventory control systems or their credit-card readers. You'll be amazed at the strange looks you'll get."

But back at the Software Factory, Jerry Hill—who himself was writing buggy code as recently as 1996—sees a silver lining. "We've learned a tough lesson," he says. "After January 1, 2000, technology no longer will be an unmanaged asset.

"When the dust settles," he continues, "Year 2000 will be remembered as technology's coming of age."

Y2K Field Tests Verify Equipment Accuracy

Deputy Postmaster General Michael S. Doughlin said the U.S. Postal Service is aggressively addressing the computer challenges posed by the Year 2000.

"Year 2000 is a daunting challenge, but I believe we have the structure and resources in place to successfully deal with it," Doughlin said at the monthly meeting of the Postal Service Board of Governors.

"The Year 2000 challenge is receiving very intense senior management visibility and attention at the Postal Service through biweekly reviews" Doughlin said. "As you might imagine, the Year 2000 challenge is a big one for us, given our size and geographic breadth."

He explained that the Postal Service has identified over 500 business applications, including payroll functions, address management databases and employee benefits processing, that will be impacted by Year 2000 conversion.

Recently completed field tests of postal mail processing equipment in Atlanta and Tampa were successful, Doughlin indicated. "The tests were able to verify the ability of that equipment to accurately sort mail in the Year 2000 environment," he said.

Last summer, the Postal Service's Office of the Inspector General (IG), at the request of postal management, began monitoring the ongoing progress and status of the agency's Year 2000 effort. Doughlin said the IG's reviews "have been helpful in covering all the bases."

The U.S. Postal Service, 11/98, p.1

What Can Be Done?

Very much can be done. The simple answer is that we can make our computers Y2K compliant. That means that all our equipment, hardware and software, can be adjusted so that it will work without any major drawbacks. We have already listed a number of firms who can make our computers Y2K compliant. The volume of information available is stunning. A large number of books, videos and CDs are available to help us.

Computer "Crash" Normal

Readers should also understand that virtually all computers experience periodic glitches.

In fact, at our office here in West Columbia, we experience some kind of computer problem several times a week. Sometimes, one of the MacIntosh computers we employ for design and typesetting our many publications crashes several times a day. Such type of computer problems should be considered normal.

Furthermore, the just mentioned problems are not related to Y2K. Remember, Y2K means Year 2000. It's a one-time event, it is unique and will take place January 1, 2000. To do "what can be done?" simply means to get compliant.

What Must Be Done?

Those whose systems are not Y2K compliant will experience problems which could lead to loss of income, even loss of business. While the answer to "what must be done" is simple, the process is somewhat complicated.

A complete inventory of all computers, with related hardware and software, must be considered now. A revision of all functions with related programs must have to be checked one by one.

One interesting article I read recently simply stated, "Follow the money route."

> Every business needs money to operate. Ask yourself: How does the money come in? At that point, you must start to ensure the continuity of cash flow.
>
> The American Business Press says proper action with respect to year 2000 compliance involves four steps: 1) Identify your internal systems and external vendors that will be affected. 2) Prepare a plan to verify the compliance of both. 3) Document your efforts. 4) Test your critical systems and audit the compliance of your critical vendors.
>
> Steve Gladyszewski, vice president, information services, Ziff-Davis, and chairman of the ABP MIS committee, offers some tips: "Test everything. Start with your internal systems and follow the money," he advises.
>
> "The further down the chain it gets from money, the further down it is on the priority list." As for service companies, you should have your complete plan together before you begin testing. And, he says, "Know that if you haven't been spending on technology over the past 10 years, it will cost you now. To be in compliance, you need fairly current technology."
>
> —*Folio*, 9/15/98, p.9

You Are Responsible!

Since each business is unique, you cannot delegate your problem to others or hope that Y2K will not affect your system. As a business owner or operator, you are exclusively responsible for your company.

It's like our relationship with the Internal Revenue Service: We may have an excellent accountant who takes care of all of our paperwork, particularly that relating to the IRS.

The accountant, or the accounting firm, may be winning all types of awards for excellent service but when it comes down to a conflict with the IRS, and discrepancies are uncovered in your tax-paying methods, then only one name is accountable, yours.

If you or your business have not paid the required tax, you must pay the penalty. Again, we summarize, the Y2K problem is real, it must be fixed, and it must be done now. To leave it to chance and hope that it will go away with time is extremely unrealistic and foolish. ∎

CHAPTER TWENTY-TWO

Y2K Interview With Joel Froese

S ome people claim that there are millions—even billions—of chips, hidden in awkward places, such as oil rigs under the ocean, etc.

In this chapter my son, Joel, our Manager Of Information Systems, gives us the following responses.

Q: What is a chip?
Joel: A chip is a package of integrated circuits. It can be as simple as a few transistors or complex like one of the latest microprocessors that are the brains of a modern computer. Simply put, they are the parts that make up any computer or embedded system.

Q: So what does a chip do?
Joel: That varies greatly from chip to chip. For example, radios contain chips that are nothing but analog amplifiers.

However, let's focus on digital chips (that deal with numbers) since they are the only kind that are susceptible to the Y2K bugs.

Every computer has a microprocessor (or, in simpler devices, a minicontroller.)

This is designed to follow instructions (a program) to act on data (information.)

A simple program is stored on a ROM (Read Only Memory) chip at the time of manufacture.

Most traditional computers, but not embedded systems, also have disks where more programs and data can be stored.

RAM (Random Access Memory) chips store data (and programs) on a temporary basis, if the power is turned off, the content is lost.

There are a host of other chips in a computer that are support chips to allow data to flow between these different parts and the outside world (a keyboard, mouse, printer, screen, etc.).

Q: What's the difference between a regular chip and an embedded chip?

Joel: An embedded chip simply means it is embedded in some type of equipment. Normally, one would expect to find computer chips in a computer. The term "embedded chip" is referring to the fact that it is embedded in something other than what we would traditionally think of as a computer.

Q: Would oil rigs, elevators, traffic lights, or our cars have any embedded chips?

Joel: Yes; there is very little equipment or appliances that don't have some kind of embedded computer chips. They generally replaced mechanical controls because they are more reliable, add functionality and simplify operation.

Q: Is it correct that independently, meaning if not connected to electricity, the chip is dead, meaning non-functional by itself?

Joel: That's right, a chip is totally useless unless hooked up in a circuit that, first of all, supplies it with power, but also it needs support chips and other components that allow it to run and interface with some kind of input and output to be useful.

With the a few exceptions, like a digital watch, one chip isn't a computer.

Q: What about time-based chips, do they have some kind of clock inside? In other words, does that chip have time in itself or do you put the time in?

Joel: Generally, time keeping is a function of programming, usually burned into a ROM chip.

This program interrupts the microprocessor every second (or more often) to update the time and date stored on a RAM chip.

If it is a simple timing cycle, it will count up seconds (or some arbitrary unit of time) from any given start time.

If it must interface with the outside word (namely human operators) it may be necessary to keep track of the actual date and time of day.

Then, it must also have a way for the operator to change the time and date due to daylight savings time or moving it to another location in a different time zone.

Q: How can somebody determine if an embedded chip has a Y2K bug?

Joel: First, you have to see if the equipment in question has provisions for entering the current date and time. That really eliminates a lot of systems.

Then you can either test it or request information from the manufacturer. In browsing the Internet, I have found that many companies now have a section on their web-site dedicated to information about the Y2K compliance of each of their products.

**Q: I have heard that some embedded chips have date
and time related programming although it isn't
used.**

Joel: In that case, how does the chip know when the year
2000 rolls around? We have to remember time constantly
changes, it is impossible to burn it in at the factory; you generally don't know what time zone the machine will be used
in, or if they have daylight savings time there.

You cannot rely on backup battery forever either; eventually they will go bad, so you have to have a way for the
end user to set the time and date.

If indeed an embedded chip is maintaining the time and
date without reference to the outside world and if it has a
Y2K bug, then it could very well fail anytime between now
and a hundred years from now, but not on January 1, 2000.

**Q: Let's say that an elevator has an embedded chip
that keeps record of maintenance. When the year
2000 comes, it may think it's the year 1900, which
means, it's 100 years overdue for maintenance, Will
it shut down the elevator?**

Joel: It is possible, but for maintenance purposes, it is more
likely that the computer (embedded chips) is simply counting time from the last time it was serviced.

This is a lot easier than comparing the date of last service
with the current date. Elevators, traffic lights and such may
be programmed to vary their schedule according to the time
of day and/or day of week to handle traffic flow.

Some do indeed maintain the current date to compare it
against a list of holidays, and if they have the Y2K problem,
it could cause a problem.

Generally these are add-on systems, and a failure would
simply revert it to a default schedule. Here are some direct
quotes from elevator manufacturers:

Your Elevators are Safe with Dover Elevator Controllers—
While Dover's elevator controllers are indeed "State of the
Art" microprocessor controllers, they are not programmed to
be "date aware." Because the Dover Elevator controllers are
not aware of the actual date, there will be no problem with
the year 2000 date change. This applies to Dover's complete
line of controllers including

[from www.DoverElevators.com]

Schindler Elevator Corporation Elevators and Escalators

Elevators and escalators manufactured by Schindler Elevator
Corporation, and those manufactured by the former
Westinghouse Elevator Company, don't use calendar or date
information to control their operation and thus won't be
affected by the Year 2000 calendar change. We recommend
that you check with the manufacturers of any other
elevator/escalator equipment on your properties for Year
2000 compliance.

[from www.schindlerElevators.com]

Q: Let's assume my computer won't recognize the year 2000. What will happen?

Joel: That depends. Unlike embedded chips that are programmed for only one function, a personal computer or
mainframe is an open system. It contains programming from
the computer manufacturer, the operating system (like
Windows or DOS for a personal computer or VMS for a
mainframe), and finally the user applications which can
come from numerous sources.

Starting with the system programming, if it only stores
two digits for the year, it might crash when it tries to add one
to 99 in a space only 2 digits wide. A crash is really not that
big a deal, it happens for various reasons. After restarting

the system, you would enter the correct date and time and the problem is resolved; who is to say that '00' doesn't mean 2000?

Q: Can it really be that easy?

Joel: No, of course not. A computer is quite useless without the user applications such as word processing, database management, etc.

While word processing would probably not be affected at all, more complex and custom programs are where the problem can be.

But those problems don't necessarily happen exactly on January 1, 2000; some of them have already presented themselves for a couple of years and other may not present themselves until the end of a fiscal year that spans 1999 and 2000.

It all boils down to the question, "Does the program compare or calculate dates, and will these dates span between the 20th and 21st century?" Under these conditions, a program with a Y2K problem could either give erroneous information, or crash.

It may be possible to try the program again and input a different date range; for example you might have to run two reports instead of one.

Q: Someone wrote that if not fixed, the computer is going to blow up or get stuck.

Joel: First of all, it is a well-known maxim that software cannot break hardware. Every new computer user is assured "Go ahead and try it, you can't break it," so there is no issue of a computer actually blowing up.

If the computer hangs up or crashes, this is nothing uncommon.

You reboot your computer and continue working.

Q: What does "reboot" mean?

Joel: This means hitting a reset button or turning the computer off and then back on again. Computers crash for a variety of reasons so most computer users are familiar with this procedure; you may lose some recent information, so it is an inconvenience, but it not unheard of.

Q: What have you done at this point to be Y2K compliant?

Joel: I tested all our computers, the phone system, fax machine, and even the postage machine by setting the date ahead to the year 2000.

I checked to see if indeed it stores the date, if the computer and programs still work, and also if it recognizes that the year 2000 is a leap year. Although this was a cursory test, I found no critical problems.

However, I will certainly check the date and time on each machine on Saturday, January 1, 2000 and correct any as necessary.

As far as our user applications go, I am confident the commercial applications we have purchased don't have any major problems.

More importantly, I have been maintaining the custom programs that I have written as potential problems arise. Since they are written in a "dBase" language, all dates are automatically stored with a 4 –digit year.

In 1991 I had to change the definition of a life subscriber from 12/1/1999 to 1/01/2059 as regular subscription began to approach this date.

I also set an epoch of 1960 in most of my programs. This means if the user enters 2 digits of a year between 60 and 99, a 19 is added to the front. If numbers between 00 and 59 are entered, a 20 is added to the front. In other places I have changed the input fields to require all 4 digits of the year (such as birth dates); the users don't like having to key in two extra digits, but I predict this will become more common in the new Millennium.

Q: If you can determine the date on your computer, all you have to do is put in the new time and date, is that correct?

Joel: Yes, you can tell the computer it's the year 2000, or year 00 as the case may be, anytime you want to. As a matter of fact Nova Scotia [Canada] Power has permanently advance the date on all their equipment to the year 2000 so there will be no surprises on new year's day.

Q: Shall we do a test right now? Let's set the date on my Casio watch to 12–31–99, 11:59 p.m. The result? It jumped right back to January 1999. The watch didn't go beyond 1999.

Joel: That's a Y2K bug all right. But what we can do now is set the date to, January 1, 2000. See, it works, your watch now reads January 2, 2000.

Q: You have read many of the articles that have come to our office, alarming the reader about year 2000. Are there any items which you legitimately believe that we have a reason to worry about?

Joel: As a business owner or even an employee it is your responsibility to make sure your business will continue. This is a serious problem that has to be addressed.

However, as an average citizen, and as a consumer, I don't see any useful precautions you need to take.

If you want to keep your bank statement, which I do any way, that's fine; but if your bank has any problems they are going to go back to their backup data, which is securely stored—there is virtually no way they can lose this.

As far as stocking up on food, water, fuel and cash, I have no problem with it if it causes you no inconvenience, but what are you going to do with a year's worth of dried beans when nothing happens?

Q: Let's say there is John Doe who operates a mail order business. Years ago his nephew wrote a program; now comes the year 2000. What would be the worst-case scenario?

Joel: If he relies on this program and it has a Y2K problem and he cannot find a work-around solution, he is indeed in deep trouble.

The problem is if this person is still in programming, he is probably busy elsewhere fixing someone else's problem. Information technology people are going to be stretched very thin for at least the first week of 2000.

This is the insidious nature of the Y2K bug, it will be happening to nearly everyone at the same time.

Due to the advances in computer technology in the last few years, John Doe may be able to buy new computers and off-the-self software at a local computer store that could do the same job.

However, although off-the-self software is versatile, it may not be as useful and efficient as custom software. There is still going to have to be some kind of data conversion, training, etc. ∎

CHAPTER TWENTY-THREE

Understanding the Y2K Problem

—by Joel Froese

W hat is the Y2K bug? First and foremost, the Y2K bug is a time-keeping problem; regardless of how complex or simple the computer or embedded chip is, it is generally immune from any Y2K problems if has no time & date clock.

Actually the term "the Y2K bug" is a misnomer, there is no one unified Y2K bug, rather a range of problems that differ in significance, scope, and reparability. The often used term "Millennium bug" is also misleading, this problem would have happened at the end of any century; it just so happened that the first practical computer was invented in the 20th century (1900's.)

Specifically the Y2K bug is a computer problem that may affect some computers and software that only store and only deal with the last two digits of the year in a date field. If a computer only stores "99" instead of 1999, there could be some problems when it compares it to a date in the year 2000 since the only thing it sees is "00." When we humans see only the last two digits, we generally know to add a "19" in front high numbers and "20" in front of low numbers such as "00" or "05." However if a computer program has not been specifically programmed to recognize this rule, it assumes that all dates are in the same century.

There are actually two distinct problems. First there is the rollover question; namely will the computer (or anything else with a computer-based time/date clock) show the correct date after midnight December 31, 1999? The second problem is the date math problem; will the computer correctly compare or calculate the difference between two dates across the century mark?

The Date Math Problem

The date math problem has already presented itself for several years now. If only 2 digits of the year are stored, expiration dates, maturity dates, etc. that extend into the year 2000 or beyond would be considered to be earlier than the current date (since 00 is less than 99.)

Also, 00 would be considered to be negative 99 years away instead of only 1 year, which could cause big miscalculations. Notice I didn't say the computer will think it is 1900 (this is a misconception); if a computer program stores only two digits, it only knows years 00 through 99.

After December 31, 1999, "00" will indeed mean 2000, and the problem computer or program will "think" that "99" means 2099.

Therefore if you were to begin using computers and software with this problem in the middle of the 21st century, you generally wouldn't have any problems until 2099. The problem occurs only when comparing dates across the century mark.

Subscription Dates

Now that many magazine expiration dates are valid well into the 21st century it is obvious these companies have dealt with this problem, otherwise you would no longer get your magazine, to the computer your subscription would have been expired over 90 years ago.

You can check the expiration date on the mailing label of practically any magazine you now receive; you will notice that even if the entire expiration date is abbreviated to 4 or 5 characters, because of limited space on the label, the magazine continues to come although expiration date show the year as 00 or 01.

Credit Cards

The same goes for credit cards; during the end of 1997 there were problems with some credit acceptance terminals, but now those have mostly been resolved. There are now thousands if not millions of people who use their credit-cards that expire in the year 2000 or beyond.

The Scope Of the Problem

Of course solving one particular problem doesn't necessarily mean the entire organization is Y2K ready. Furthermore, not all organizations have dealt with this problem yet (some smaller, less technically savvy companies may not even have discovered it yet.)

They will have several choices come January 2000, either close out all "historical" data so that all new dates will fall in the 21st, or when running report, labels, etc. they have to run one batch from whatever beginning date, to December 31, 1999 then run another batch from January 1, 2000 to whatever ending date.

This, of course, isn't always feasible, so there will no doubt be mad scramble to buy new computers and/or off the shelf software to replace these custom systems. This could translate into temporary increase in staffing that manual and less efficient processes require (which could be a problem due to current record low unemployment.)

The Fix
We have all heard the horror stories about the effort required to fix Y2K bugs, namely combing through thousands or even millions of lines of code looking for elusive date references. This, however, is only the worst case scenario for poorly developed and poorly document programming.

To make a change such as increasing the year field to 4 digits should be a simple task in an well-organized program. Programming magazines are full of ads for tools that assist in or even totally automate this process.

When using a modern programming language that has a built in date field, no changes may be necessary in the source code at all, just a recompile (a compiler changes human-readable source code into machine language that the computer uses.) with a new compiler.

Who Will Be Affected?
We always assume that every organization will be affected by the Y2K problem. This isn't necessarily true; by using ordinary personal computers and off-the-shelf software it is very likely that an organization will have no problems at all with computers on January 1, 2000. Even on older PC's that don't rollover correctly, the solution is relatively easy: just correct the date once and you're back in business.

While Y2K-related bugs have been reported in modern software, the patches (a patch is a slightly newer version of any given program) are usually available immediately thanks to the Internet.

If the software isn't "patched" it still isn't necessarily going to be a problem for everyone because a "bug" by definition is a problem with software that causes an incorrect response due to an unusual condition. (Therefore the term "Y2K bug" is actually a misnomer, the year 2000 isn't unusual, it is inevitable!)

For example, the well-publicized discrepancies between Microsoft's word processing, spreadsheet, and database programs handling of dates with 2–digit years can easily be avoided by typing all four digits for every date, every time. In many cases programs with real Y2K problems will work fine because the dates stored are insignificant or the program is only used for simple tasks and the problem only crops up when specific features are used.

Rollover Problem

The rollover problem could affect more computers and microprocessor controlled machines. However, it isn't impossible to determine what will be affected. First of all, we must understand that in order for anything to have the Y2K rollover problem it must have the following:

It must have a real-time (and date) clock (RTC) that runs all the time regardless of what the rest of the computer or machine does

This RTC must have a backup battery that keeps the clock running even when the power is shut off.

If it doesn't have a battery or even if it does and the battery goes dead, there must be some way of setting the date (and time) by the user, which means it must have some kind of readout to display the date and/or time.

And of course, for there to be any problem, it must store only two digits for the year.

This really limits the affected machines; if you think about it, your VCR and a personal computer or electronic organizer (if you have one,) are the only things in your home that you set the time and date on. Programmable coffee makers, alarm clock radios, etc. are totally immune from any Y2K problems since they only keep time.

The rollover problem could present itself in several different ways depending on the original design. When you add one to 99 you get 100, but if there is only room for two digits, you have created an overflow error (just as if you add one to 99,999,999 on an eight-digit calculator.) This could "crash" the computer or machine.

The term "crash" simply means the computer stops working. Anyone who has used a computer for more than a few hours is probably familiar with this situation. The next step is to "re-boot" the computer by pressing a reset button or by turning the power off and then back on.

Since the RTC runs on a backup battery, it may necessary to also temporarily remove the backup battery. When the machine is turned back on the date will be January 1, 00 or some other default year such as 70 or 80; after the correct time and date is set, the problem is most likely over. Setting the time and date on any machine is normally easy, in most cases it is done at least twice a year (when the time changes for daylight savings time) anyhow.

This is a worst-case scenario, more likely the date will simple rollover to 00 as it should and if the date is only used for display or logging, there is no problem at all. However, there may still be some date-math problems if the current date is compared to previous ones as discussed previously. But it is important to remember that the rollover problem is generally quick and easy to remedy, therefore the only real problem is the date-math problem.

Personal Computers and Y2K

Personal computers and PC based networks, which have nearly replaced mainframe computers and terminals in many businesses, were developed in the early 1980's. From the very beginning the system date had a 4-digit year. The first real database program for PC's, dBase III, also stored dates with a 4-digit year (although by default it only displayed two digits.)

Since it had no real time clock, every time you started an original IBM personal computer it would first ask you for the date and time, which you could enter with or without the extra two digits in the year (it would automatically add "19" if you didn't.) It soon became apparent that this was a nuisance since, unlike mainframes, personal computers were turned on and off at least once a day.

The Real Time Clock (RTC)

Then came the real time clock (RTC), first as a third-party add-on and later built-in on new computers. Usually it is built right into the motherboard (the heart of any computer; all other accessories are then connected to the motherboard,) the RTC is a separate circuit that runs continuously regardless of what the computer is doing.

Even when the computer is turned off and unplugged, it runs off a small battery that you find in every modern personal computer.

When you turn a personal computer on, one of the first thing that happens is the computer retrieves the time and date from the RTC; after that, the operating system (such as DOS or Windows 98) maintains a "running clock" in memory.

Therefore, from the system date standpoint, all personal computers will sail over the Millennium mark without a hitch if they are up and running on New Year's eve.

However, due to inconsistencies by computer parts manufactures, the RTC may be storing the wrong date.

This will make itself apparent when the computer is turned on for the first time in the new Millennium.

Of course, the solution is to simply change the date (which affects both the RTC and operating system's "running clock.") After that the RTC will maintain the correct date and time.

Test Your PC Today

If you have a personal computer you can test this. Set the date to 12–31–1999 and the time to 23:59:45; you will quickly see that the operating system has indeed correctly rolled over to the new Millennium. Now turn the computer off for a few seconds, then back on; the date may now be incorrect on some older computers, but if you correct it (set it to 01–01–2000) and cycle the power again the date should be correct.

Apple Computers

Apple's Macintosh computers were designed from their inception with a very sophisticated time & date field that includes not only the century (4 digit year) but even the time zone of the location of the computer, which can be useful for comparing creation/modification dates of documents from around the globe. And since only Apple Inc. made the hardware, there are less of the inconsistencies that plague some PC's.

Macintosh computers are like a gold standard when it comes to Y2K compatibility, however it doesn't prevent programmers who write the user applications from using their own method of storing dates. As a matter of fact, a few Y2K problems have already been reported in Macintosh programs; this is part of the insidious nature of the Y2K problem.

Scheduled Service Shutdown

Much has been made of the fact that some machines are programmed to shut down if it has not been serviced in a certain amount of time. Assuming these machines also have a Y2K problem, some people expect all kinds of manufacturing machinery, elevators, power transmission equipment, and etc. to grind to a halt on January 1, 2000.

However, if the machine simply keeps track of the time between service, then there is no reason to keep track of the date and time since this adds a lot of complexity. As we all know computers deal only with number (binary number specifically) and all other data is just a layer of abstraction added to computers for the sake of the operators and programmers (humans.) For example, if the machine must be serviced at least once a year the computer can be programmed to shut down the machine after it counts 31,536,000 seconds (or some other arbitrary unit of time) since the last time a certain switch was pushed.

Even though that seem like a big number to us, since it is a numeric value it take only 25 bits (binary digits) to store and no extra programming (all computer microprocessors can increment a number with only one instruction.) To do this with a standard date and time field (such as 12/10/99 23:45:01) would take at least 96 bits and lot of extra programming (to increment minutes after 60 seconds, hours after 60 minutes, day after 24 hours, and etc. plus the date math to compare it with the last service date.) Unless there is a screen or printout with a human-readable date, the computer doesn't deal with dates at all and therefore is immune to any Y2K problems.

Elevators
Otis (the worlds largest maker of elevators) has refuted a now ubiquitous urban legend of elevators being stuck between floors as a result, of a Y2K bug for some time now:

So what do we mean by Year 2000 compliant? Simply stated, if you own an Otis manufactured product serviced by Otis, the product will continue to operate on January 1, 2000, just as it did on December 31, 1999. This includes Otis' Remote Elevator Monitoring, or REM III, product.

(www.otis.com)

Airlines

An article in the January 21, 1999 issue of *Travel Weekly* (a magazine for travel agents) addressed the concerns of the flying public.

The two major aircraft manufactures, Boeing and Airbus Industries, set self-imposed deadlines for Y2K compliance a year ahead of time (Boeing, Dec 31, 1998; Airbus, the next day). ... Airbus said it detected no issues affecting overall airworthiness or aircraft operations. ... Boeing ... said "no safety-of-flight issues related to Y2K for airborne systems exist for an airplane in flight."

This isn't surprising, since aircraft systems are thoroughly tested for every conceivable situation due to the unforgiving nature of a malfunction in flight.

Aircraft designers must consider all kinds of unusual condition. For example, an aircraft's altimeter will report a "0" then negative numbers when landing in Amsterdam.

This information is passed on to other devices in the cockpit, which could cause a "divide by zero" error if the designers were not as careful as they are.

This would only be discovered when the aircraft dips below sea level at one of the few places on Earth where that is possible (such as Death Valley, CA or near the Dead Sea in Israel.)

Rest assured these designers examine how their equipment deals with all types of scenarios no matter how unusual or unlikely. This also applies to most other critical devices such as medical equipment.

The Ticket Counter

Another article on the same page examined several airlines' Y2K readiness. Since tickets for travel in January 2000 can be sold 11 months in advance, February 4, 1999 was the deadline for having their reservation systems fixed.

Although these systems are not as critical as an aircraft's systems, they are very critical to the financial health of the airline; the same goes for practically any other kind of corporation for that matter!

The FAA has antiquated air traffic control computers that have already caused their share of problems over the last few years (including shutdowns of regional radar centers.)

Air traffic control is actually not a very automated process.

The radarscopes themselves can operate without computers and the majority of air traffic controllers were trained how to track aircraft manually.

If these computers have Y2K problems, there won't necessarily be any immediate danger, but air traffic may be delayed which is nothing new to experience air travelers.

2000 Is a Leap Year

There has been some concern that we are still not "home free" if everything still works on January 2000; there still may be a problem at the end of February because it is leap year. We all know that a leap year is any year that is divisible by 4, however there are two exceptions: a) if the year is divisible by 100, it isn't a leap year b) unless it is also divisible by 400 (like 1600, 2000 & 2400.)

While it is conceivable that some programmers may have simply assumed all years ending in 00 are not leap years (since 1900 wasn't a leap year,) it is more likely that if shortcuts were taken in some programs, the assumption was made that all years divisible by 4 are leap years, in which case this will work fine until 2100.

If there really is a problem discovered on Wednesday, February 29, 2000, a solution could be as simple as closing the business for that day. Then on Thursday correct the date on all computers and resume work.

The "Old" Y2K Problem

The Y2K problem isn't new; when computers came into widespread use in the 1960's and 70's, there were far more people still alive who had been born in the 19th century than there are today.

No doubt problems arose and were dealt with when calculations were made based on their birth date.

Of course most of those systems are no longer being used and replacement systems may have actually re-introduced a Y2K problem that was already corrected on the old system.

A perfect example of this is related in a recent issue of TIME magazine:

> "... a group of Mormons in the late '50s who wanted to enlist the newfangled machines [computers] in their massive genealogy project—clearly the kind of work that calls for thinking outside the 20th century box. [IBM COBOL developer Robert] Bemer obliged by inventing the picture clause, which allowed for a four-digit year.
>
> From this point on, more than 40 years ahead of schedule, the technology was available for every computer in the world to become Y2K compliant."
>
> —TIME, January 18, 1999

Although this particular programming language apparently required the "invention" of a picture clause, almost any programmer could have dealt with this on his own because early programming languages didn't necessarily have a standardized way to store dates, they only had numeric (for numbers) and character (for letters—text) data types. My point is that this is a known issue and far-sighted programmers (more than we may think) have been addressing this for decades.

No Mystery Here

It is important to remember that Y2K is a computer problem and as such the extent of the problem can be precisely determined; it isn't a mystery.

That is why large corporations are doing in-depth Y2K inventories of all computers, software, and other machines that may be affected. This extends all the way up to large-scale, industry-wide Y2K readiness testing like what the securities investment community has already done.

Nevertheless, we know fixing one problem doesn't automatically mean everything is fixed.

There could still be problems in the very same program not to mention the same computer or even other systems; there is no "magic bullet." To solve this problem requires careful, thorough, and sometimes tedious work, but it isn't impossible.

Computer problems are nothing new, they have occurred in the past, and will continue to plague us in the future as software become more and more complex.

The Y2K bug is a known problem and there are precise, analytical steps to deal with it. The following is an informative list of steps a business needs to take to be ready for the year 2000 from the National Association of Securities Dealers, Inc.:

What steps can I take to make my business Year 2000 compliant? Typically, the 6 phases that have been helpful to businesses preparing for the Year 2000 challenge are:

1. Awareness—These are the steps an organization undergoes to understand Year 2000 issues and their potential impact. This includes employee education on Year 2000 issues. (e.g., flyers, workshops, seminars), creation of reporting structures, project responsibilities,

the creation of a project to monitor and protect the organization against risk.

2. Assessment—This phase usually involves the organization's preparation for dealing with and identifying the specific risk to their organization. Typically it starts with creating a physical inventory, measuring risk, and contacting third party vendors. Products of the assessment phase might include timelines with milestones, inventory lists, resource and budget allocation.

3. Remediation—This phase is usually referred to as the "fix it" phase. The organization begins replacing any non-compliant business equipment or software, program code modification, management of vendor solutions and renovation of systems.

4. Testing—This phase is to allow the business to know that its business will operate. It involves the testing all your systems and inventory items. Best practices for this phase includes setting up a separate test environment, allowing adequate time for testing, and documenting results systematically. Types of testing include: individual testing, point-to-point testing, and industry testing.

5. Implementation—Installing new or upgraded systems and performing testing along the way.

6. Contingency Planning—This is the management plan for mitigating and/or correcting organization's exposure to risk and failures due to Year 2000 related issues. Many organizations refer to is as the guide for "what to do when things don't go as planned."

http://www.nasd.com

Electrical Power Plants

Regardless of how Y2K ready any computer is, it must have electricity to run.

Many people are worried that there are Y2K bugs in power plants and substation around the country and the world, and that if enough equipment fails the effect would cascade across the power grid and cause a nationwide or world-wide blackout.

What we must remember is that when computers are used in critical application such as this, they are usually not given complete control.

It is understood that such a complex and fragile machine may crash at anytime.

That is one of the reasons why you see a control room with hard-wired switches, lights, and mechanical dials at any power plant.

A computer could easily monitor all these inputs and take appropriate actions without a single operator, but in an emergency, a living, thinking human has to be in control to handle the myriad possibilities that could arise.

A competent operator can develop a work-around for broken equipment or even multiple problems.

At Work On New Year's Eve

Any machine, no matter how sophisticated, can only be programmed to handle a finite number of possibilities. Since this weakness and our dependency on electricity is known and the year 2000 now looms so close, doubtless power companies around the country and the world are reviewing and practicing their manual procedure.

Unfortunately for most utility workers, they will probably have to be at work on New Year's eve. Additionally anyone in a computer trade will probably be awake somewhere baby-sitting a computer regardless of whether it needs it because this problem has been publicized to the point of hype.

We must remember that engineers, programmers, and inventors of this century developed all of our modern technology and many of those people are still around. More importantly, this has developed into an intellectual infrastructure that trains new technology workers with their cumulative knowledge plus the very latest technology that allow them to develop even more efficient and sophisticated machines.

Some people revel in the irony that modern technology could destroy itself and not even our brightest mind can save it. However there isn't enough evidence to support this wild speculation.

This technology wasn't dropped from outer space, as some UFO cultist may believe. There really are people who understand it and have been working on this problem for some time now.

Spiritual and Daily Living Lesson
The Y2K bug a perfect example of the end result of procrastination, since this problem has been known about for a long time.

It is easy to excuse the programmers and management that allowed this to happen back in the 1960s and '70s; very little of the hardware or software from that era is still being used.

However, in the intervening 30–40 years these systems have most certainly been totally overhaul or re-engineered several times when moved to new computers, databases, and/or programming languages.

This was the perfect time to address the Y2K issue, (and to be fair, many people did) but since there was no immediate benefit to fixing it since the year 2000 seemed so far away, it wasn't done. Again and again the opportunity presented itself but was pushed aside by supposedly more important matters.

Now, all of a sudden there isn't enough time left, and a lot of money and time has to be spent to fix something that could have been just a sideline on another project.

The lesson for your personal life is obvious! ■

Order Extra Copies of this book for friends, relatives and co-workers.

Mail to: The Olive Press • P.O. Box 280008 • Columbia, SC 29228

❏ Please RUSH the book "When Y2K Dies!"

❏ 1 Copy $9.99   ❏ 2 Copies $15   ❏ 3 Copies $20

Postage and Handling $2.45 for any order

Total Enclosed: $ _____

❏ Check   ❏ Cash   ❏ Credit Card

#_____

Exp_____ Phone_____

Name_____

Address_____

City_____

State_____ Zip_____

1272

⫸ *Tap into the Bible analysis of top prophecy authorities...*

Midnight Call is a hard-hitting Bible-based magazine loaded with news, commentary, special features, and teaching, illustrated with explosive color pictures and graphics. Join hundreds of thousands of readers in 140 countries who enjoy this magazine regularly!

⫸ *The world's leading prophetic Bible magazine*

⫸ *Covering international topics with detailed commentary*

⫸ *Bold, uncompromising Biblical stands on issues*

⫸ *Pro-family, Pro-life, Pro-Bible*

12 issues/1 yr. $28.95
24 issues/2 yr. $45

1272

Mail with payment to: Midnight Call • P.O. Box 280008 • Columbia, SC 29228
❑ YES! I would like to subscribe to Midnight Call magazine!
      ❑ 12 issues/1 yr. $28⁹⁵ ❑ 24 issues/2 yr. $45
      ❑ Cash   ❑ Check   ❑ Master/Visa/Discover
      With Credit Card, you may also dial toll-free, 1–800–845–2420

Card#_____Exp:_____

Name:_____

Address:_____

City:_____ St:_____ Zip:_____